HERO TALES OF OUR NAVY

BY
GEORGE GIBBS

Introduction by
Graeme R. Newman

READ-ME.ORG INC.
OPEN ACCESS PUBLISHERS
READ ME .ORG
Australia, New York & Philadelphia

Introduction

By Graeme R. Newman

"Hero Tales of Our Navy" by George Gibbs, also known as "Pike & Cutlass: Hero Tales of Our Navy," is a captivating collection of stories that celebrate the bravery and adventurous spirit of American naval heroes. Written in the late 19th century, this book provides a historical account of significant naval battles and the daring exploits of notable figures in the early history of the United States Navy.

The book opens with the audacious actions of John Paul Jones, a prominent figure in American naval history. In April 1778, commanding the sloop "Ranger," Jones embarks on a mission to disrupt British shipping. His daring plan includes an attack on the town of Whitehaven, which he successfully infiltrates with a small crew. This initial story sets the tone for the book, highlighting Jones's exceptional leadership and tactical prowess.

Throughout the book, Gibbs recounts various thrilling episodes that showcase the courage and ingenuity of American sailors. These tales are not just about battles and victories but also about the personal valor and resilience of the individuals involved. The stories span different periods and conflicts, providing a comprehensive look at the evolution of the U.S. Navy and its role in shaping the nation's history.

Gibbs's writing style is engaging and vivid, bringing to life the intense and often perilous experiences of these naval heroes. His detailed descriptions and narrative flair make the historical events accessible and exciting for readers. The book serves as both an educational resource and an inspiring tribute to the men who risked their lives for their country.

"Hero Tales of Our Navy" is more than just a collection of war stories; it is a celebration of the spirit of adventure, bravery, and patriotism that defines the American naval tradition. Whether you are a history enthusiast or simply enjoy tales of heroism, this book offers a fascinating glimpse into the lives and deeds of some of the most remarkable figures in naval history.

The book had a notable impact on popular culture, particularly in the late 19th and early 20th centuries. The ways in which it influenced society may be summarized thus:

1. **Patriotic Inspiration**: The book's vivid recounting of naval heroism inspired a sense of patriotism and national pride. By highlighting the bravery and strategic brilliance of American naval officers, it reinforced the idea of the United States as a formidable maritime power.
2. **Educational Value**: The stories served as educational material, providing readers with historical knowledge about significant naval battles and figures. This helped to popularize naval

history and made it accessible to a broader audience, including young readers who might have been inspired to pursue careers in the navy.

3. **Cultural Representation**: The tales contributed to the cultural narrative of American exceptionalism and the valorization of military service. They helped shape the public's perception of the navy and its role in protecting and advancing national interests.
4. **Literary Influence**: Gibbs's engaging storytelling style influenced other writers and historians. His ability to blend factual history with compelling narratives set a standard for how military history could be presented in an entertaining yet informative manner.
5. **Media Adaptations**: While there are no direct film or television adaptations of Gibbs's book, the themes and stories of naval heroism have been echoed in various forms of media, including movies, TV shows, and other literary works. The book's influence can be seen in the broader genre of naval adventure stories that followed.

Overall, "Hero Tales of Our Navy" played a significant role in shaping the cultural and historical understanding of the U.S. Navy's early exploits, leaving a lasting legacy in American popular culture. Overall, the heroism of naval figures during this period not only shaped military history but also left a lasting legacy in the arts and literature, inspiring generations with tales of courage and adventure.

Major themes and their insemination channels include:

- Art expressing romanticism and realism, via public monuments and memorials, illustrations and prints.
- Literature, including historical narratives, patriotic poetry
- Culture and national identity
- Popular media and heroism

HERO TALES OF OUR NAVY

PIKE & CUTLASS

WRITTEN & ILLUSTRATED
BY GEORGE GIBBS

THIRD EDITION, ENLARGED

PHILADELPHIA & LONDON
J. B. LIPPINCOTT COMPANY
1925

THE ESCAPE OF THE "CONSTITUTION"

PRINTED BY J. B. LIPPINCOTT COMPANY
AT THE WASHINGTON SQUARE PRESS
PHILADELPHIA, U. S. A.

TO THE CADETS

OF THE

UNITED STATES NAVAL ACADEMY

PAST, PRESENT, AND

FUTURE

CONTENTS

LIST OF ILLUSTRATIONS

THE EFFRONTERY OF PAUL JONES

IN April, 1778, there were more than two-score of French ships-of-the-line within easy sailing distance of the coast of England. They were tremendous three-decked monsters, armed with tier upon tier of cannon, and it took nearly a thousand officers and men to man each of them. They lay at anchor in the harbors of France or sallied forth into the open sea to the southward to prey upon the commerce of Great Britain. But grand as they were, not one of them dared to do what John Paul Jones did in the little Continental sloop of war "Ranger." By good seamanship, an element of chance, and a reckless daring almost without precedent, he accomplished under the very noses of the gold-laced French admirals what they had been hemming and hawing about since the beginning of the war.

Inaction weighed upon the mind of Paul Jones more heavily than the hardest of labor. He had to be up and doing all the time, or trouble was brewing for everybody on shipboard. So when he reached Nantes, France, and found that the frigate which had been promised him was not forthcoming, he determined, alone and unaided, to do with the little "Ranger" what he was not yet destined to do with a bigger ship. No person but Paul Jones would for a moment have considered such a desperate project as the one he conceived. What the flower of the navy and chivalry of France had refused to attempt was little short of suicide for the mad American. But Jones was not cast in an ordinary mould. When he got to Brest, he made up his mind once and for all, by one good fire of British shipping to put an end to all the ship and town burnings in America.

There was clanking of bit and chain as the anchor was hove up short on the little craft. The officers and men of the great vessels of the French fleet looked over the glistening water, warmed by the afternoon sun of spring, and wondered where their impetuous harbor-mate was off to. A week before, they knew Paul Jones had demanded that the French Admiral salute the Continental flag which the "Ranger" wore for the first time. And they had given those salutes right willingly, acknowledging publicly the nation they had been helping in secret. They knew he was a man of determination, and they wondered what the American was going to do. Some of them—the younger ones—wished they too were aboard the dainty little craft, bound out to sea under a man who feared nothing and dared everything. They heard the whistles and hoarse calls of the bos'n as the men tumbled down from aloft, the sheets flew home, and yards went up to their blocks with a clatter and a rush that showed how willing were the hands at the tackles. The tops'ls caught a fine breeze from the southward and, bracing up, the "Ranger" flew down the harbor and around the point of Quiberon just as the sun was setting behind the purple cloud-streaks along the line of limitless ocean. Up the coast she moved, her bowsprit pointing fearlessly to the north, where lay the Scilly Isles. The Frenchmen left behind in the harbor looked enviously at the patch of gold, growing every moment more indistinct in the fading light, and said "En voilà un brave!"

The next day Jones left the Scilly Isles on his starboard quarter and steered boldly up Saint George's Channel into the wide Irish Sea. The merchantmen he boarded and captured or scuttled did not quite know what to make of a man who feared so little that he looked into the eyes of the lion sternly and even menacingly when one movement might have destroyed him. These channel-men thought themselves secure, for such a venturesome procedure as that of Paul Jones was contrary to all precedent. They couldn't understand it at all until their vessels were burned and they themselves were prisoners. Then they knew that they had been taken by a man whose daring far surpassed that of the naval captains of England and France. In plain sight of land he took a brig bound from Ireland to Ostend. He didn't want to be bothered with prisoners, so he sent her crew ashore in their own boat to tell the story of their escape. Then off Dublin he took another ship, the "Lord Chatham," and sent her in charge of a prize-crew down to Brest.

Paul Jones had one great advantage. Nowadays, when the railway and telegraph have brought all the people of the world closer together, such a cruise would be impossible. The report would be sent at once to the Admiralty, and two fleets, if necessary, would be despatched post-haste to intercept him. But Paul Jones knew the value of the unexpected. And although fortune favors the brave and the winds and waves seem always on the side of the ablest navigators, he had made his calculations carefully. He knew that unless an English fleet was at some point nearer than Portsmouth he would have ample time to carry out his plans.

He made up his mind before burning any shipping to capture, if possible, the Earl of Selkirk, who lived on St. Mary's Isle, and to hold him as a hostage. By this means he hoped to compel England to treat American prisoners with humanity, according to the laws of war. But on the twenty-first of April he picked up a fisherman who gave him information which for the moment drove all thought of the Earl of Selkirk and the shipping from his mind. Inside the harbor of Carrickfergus, where Belfast is, lay a man-of-war of twenty guns, the "Drake," a large ship, with more men than the "Ranger" carried. He would drop down alongside of her under cover of the night and board her before her crew could tumble out of their hammocks. Such an attempt in a fortified harbor of the enemy would not have occurred to most men, but Paul Jones believed in achieving the impossible. He waited until nightfall, and then, with a wind freshening almost to a gale, sped up the harbor. The "Drake" lay well out in the roadstead, her anchor lights only marking her position in the blackness of the night. Carefully watching his time, Captain Jones stood forward looking at the lights that showed how she swung to the tide. He kept full headway on the "Ranger," until she could swing up into the wind almost under the jib-boom of the Englishman. By dropping his anchor across the chain of the "Drake" he hoped to swing down alongside, grapple, and board before the crew were fairly awake.

But this time he was destined to fail. Everything depended on the dropping of the anchor at the proper time. His orders were not obeyed, for not until the "Ranger" had drifted clear of the Englishman's chain did the splash come. Then it was too late. Fortunately the watch on the "Drake" were not suspicious. Had they been wider awake they would have had the "Ranger" at their mercy, and Paul Jones might not have survived to fight them a few days later. As it was, they only swore at the stupidity of the Irish lubber they thought he was. Jones knew that his chance was gone, and as soon as a strain came on the cable it was cut, and he filled away to sea again.

He now returned to his original plan of burning the shipping of some important town. He decided on Whitehaven as his first objective point, and the "Ranger," sailing leisurely over, dropped anchor in the outer harbor during the following night.

THE EFFRONTERY OF PAUL JONES

Whitehaven was a town of considerable importance in the Scottish and North of England shipping trade. The inhabitants were for the greater part sailors and others who made their living by the sea, and there was never a time when the docks were not crowded with vessels, of all countries, from the sloop to the full-rigged ship, discharging or taking on cargoes which figured largely in England's commerce. At one side of the harbor lay the town, and farther around to the left lay the docks where the shipping was. Over two hundred vessels, large and small, lay there or out in the roadstead. Two forts, mounting fifteen guns each, guarded the town. They were adequately garrisoned, and it looked like a piece of desperate folly to make the attempt upon a town directly under their guns.

Paul Jones knew Whitehaven from his childhood. He remembered just where the guardhouses were to be found, and knew how to force the entrance to the barracks. By three o'clock in the morning he was ready to make the assault. Two cutters with fifteen men in each, armed with cutlasses and pistols, were all he took to do the work. With thirty men he went fearlessly and confidently to intimidate the soldiers, spike the guns in the forts, overawe the town, and burn the shipping! Lieutenant Wallingford was given command of one of the cutters. His mission was to burn the shipping to the left. The other cutter Paul Jones commanded himself, and assumed the more hazardous duty of holding with his fifteen men the forts and the town, until such a blaze should illumine the morning sky that all England would know that the burning of Portland, Maine, was avenged.

Quietly they pulled up towards the great stone dock, where the shipping-houses were. The tide was very low as they moved past the schooners and brigs in the harbor, many of them careened far over on their sides, waiting for a rise in the tide to pull down to more comfortable moorings. But the boats went by without challenge or notice, and Wallingford's cutter had slipped away like a gray shadow in the darkness. The first violet streaks of dawn were just beginning to throw the shore-line to the east in hazy silhouette when they reached the landing-place.

The dawn was coming up quickly now, and Paul Jones led his fifteen men at a run to the nearest fort. With cutlass in one hand and pistol in the other, they dashed upon the first sentry. There was no time for stealth, so they bore him down by sheer weight. The next one saw them coming, but Jones locked him and the rest of them in the guard-house. Then he proceeded to spike the guns. So quick was the work that not a shot was fired. They were running towards the second fort before the soldiers were quite sure what had happened. Even then they were too terrified to follow in pursuit. As the gallant band ran towards the other fort they got a clear view of the harbor, a glimmering sheet of orange and violet, under the morning glow. But strain his eyes as he might, their captain could get no sign of Wallingford or his work. They dashed as desperately at this fort as at the other and were equally successful, intimidating the garrison and spiking every gun they could find.

But what could be the trouble with Wallingford? Still seeing no blaze or even spark among the shipping to the eastward, Paul Jones felt that the main object of his descent upon the town was to prove a failure. So he dashed down the street from the fort towards the dock, pistol in hand, followed by his crew, who rolled along grinning at the ease with which they had accomplished their work. One of them had a bad cut over the head and the blood was staining his shoulder, but he didn't seem to mind it in the least. To their surprise as they passed the houses the people began coming out of their doors shaking their fists at and cursing them. They grinned no longer, for they knew that some one had betrayed them. Jones looked around for the fifteenth man. The fellow with the cut wiped some blood from his cheek and said,—

"Dave Freeman, sir, he's gone!"

Freeman was the traitor, then.

But there was no time for parley or revenge. The mob was collecting in the street they had left and soon would be down on the dock. Though Wallingford failed, Paul Jones would not. He dashed into a house on the dock, and seizing a burning brand went aboard one of the largest vessels of the fleet. He hastily pulled together some straw and hatchway gratings and soon had a roaring blaze. Then one of his men spilled a barrel of tar in the midst of it to make the destruction more sure.

He had been so intent upon his work that he had not noticed the mob that had gathered on the dock. The place seemed black with people, and their number was increasing every minute. Then, leaving the work of destruction to the others, he went down alone to face fifteen hundred infuriated people with a single flint-lock pistol! Dave Freeman had done his work well, for they seemed to pour from every street and doorway. But Paul Jones was de-

THE DESCENT ON WHITEHAVEN

termined that the work should be finished, and took a position where he could command the boat-landing and retreat of his men. The people came down in a body to within twenty paces of Paul Jones and then—stopped. There was something in the *look* of the man and the menacing black barrel that moved from one to the other that made them quail and fall over each other to get out of range. Those in the background swore and pushed gallantly, but the front rank was a line of straw, and Paul Jones moved it with his old flint-lock as though a Biscay wind-squall was striking it. For fifteen minutes and longer he stood there, immovable, the master of the situation, the picture of the intimidating power of one resolute man over a mob. Such another instance is hardly to be found in history.

When the black smoke rolled up from half a dozen vessels of the fleet, Paul Jones's crew retreated in an orderly manner to the cutter. Jones walked down the steps into the boat, covering the crowd the while. Then his men leisurely rowed away, not a shot having been fired. It was not until the cutter was well out into the bay that some of the bewildered soldiers recovered sufficiently to load two cannon that Paul Jones had overlooked.

These they brought to bear upon the cutter dancing down in the sunrise towards the "Ranger" and fired. The shot whistled wide of the mark, and Jones, to show his contempt of such long-range courage, fired only his pistol in return.

But that was not the end of this remarkable cruise. Having failed to find the Earl of Selkirk on St. Mary's Isle, Paul Jones squared away to the southward, hoping to pick up another full-rigged ship off Dublin or to meet with the "Drake" again. He knew that by this time the Admiralty was well informed as to his whereabouts, and that before many hours had passed he would be obliged to run the gauntlet of a whole line of British fire. But he hated to be beaten at anything, and since the night when he failed to grapple her had been burning to try conclusions yard-arm to yard-arm with the "Drake."

On the twenty-fourth of April, just two weeks after sailing from the harbor of Brest, he hove to off the Lough of Belfast, where within the harbor he could plainly see the tall spars of the Englishman swinging at his anchorage. Paul Jones was puzzled at first to know how he was to lure the "Drake" out to sea, for a battle under the lee of the land in the harbor was not to be thought of. So he went about from one tack to another, wearing ship and backing and filling, until the curiosity of the English captain, Burdon, was thoroughly aroused, and he sent one of his junior officers out in a cutter to find out who the stranger was. Jones ran his guns in and manœuvred so cleverly that the stern of the "Ranger" was kept towards the boat until he was well aboard. The young officer was rather suspicious, but, nothing daunted, pulled up to the gangway in true man-o'-war style and went on deck. There he was met by an officer, who courteously informed him that he was on board the Continental sloop of war "Ranger," Captain Paul Jones, and that he and his boat's crew were prisoners of war.

In the meanwhile Captain Burdon, finding that his boat's crew did not return, got up his anchor, shook out his sails, and cleared ship for action. He was already suspicious, and too good a seaman to let unpreparedness play any part in his actions. There was not very much wind, and slowly the "Drake" bore down on the silent vessel which lay, sails flapping idly as she rolled, on the swell of the Irish Sea. As the afternoon drew on the wind almost failed, so that it was an hour before sunset before the "Drake" could get within speaking range. Hardly a ripple stirred the surface of the glassy swells, and the stillness was ominous and oppressive.

When within a cable's length of the "Ranger" Captain Burdon sent up his colors. Captain Jones followed his lead in a moment by running up the Stars and Stripes.

Suddenly a voice, looming big and hoarse in the silence, came from the "Drake,"—

"What ship is that?"

Paul Jones mounted the hammock nettings and, putting his speaking-trumpet to his lips, coolly replied,—

"The American Continental ship 'Ranger.' We have been waiting for you. The sun is but little more than an hour from setting, and it is time to begin."

Then he turned and gave a low order to the man at the wheel, and the "Ranger" wore around so that her broadside would bear. Paul Jones always believed in striking the first blow. When they came before the wind the word was passed, and a mass of flame seemed to leap clear across the intervening water to the "Drake." The "Ranger" shuddered with the shock and felt in a moment the crashing of the other's broadside through her hull and rigging. The battle was on in earnest. Yard-arm to yard-arm they went, drifting down the wind, and the deep thundering of the cannonade was carried over to the Irish hills, where masses of people were watching the smoke-enveloped duel. The sun sank low, touching the purple hilltops, a golden ball that shed a ruddy glow over the scene and made the spectacle seem a dream rather than reality. Still they fought on.

It was a glorious fight—and as fair a one as history records. The "Drake" pounded away at the "Ranger's" hull alone, while Jones was doing all he could with his smaller pieces to cripple his enemy's rigging. First the "Drake's" fore-tops'l yard was cut in two. The main dropped next, and the mizzen gaff was shot away. For purposes of manœuvring, the "Drake" was useless and drifted down, her jib trailing in the water and her shrouds and rigging dragging astern. She was almost a wreck. As she heeled over on the swell, the gunners on the "Ranger" could see human blood mingling with the water of the division tubs that came from her scuppers. The first flag was shot away, but another was quickly run up to its place. In a moment that too was shot away from the hoisting halyard and fell into the water astern, where it trailed among the wreckage. But still she fought on.

On the "Ranger" the loss had been comparatively slight. Lieutenant Wallingford and one other man had been killed and there were five or six wounded men in the cockpit. Jones seemed to be everywhere, but still remained uninjured and directed the firing until the end. He saw that the sharpshooters in his tops were doing terrific execution on the decks of his adversary, and at last he saw the imposing figure of Captain Burdon twist around for a second and then sink down to the deck. Another officer fell, and in a moment above the crash of division firing and the rattle of the musketry overhead he heard a cry for quarter.

The battle was at an end in a little over an hour. It was almost as great a victory as that of the "Bonhomme Richard" over the "Serapis." Paul Jones's ship carried eighteen guns; the Englishman carried twenty. The "Ranger" had one hundred and twenty-three men; the "Drake" had one hundred and fifty-one and carried many volunteers besides. The "Ranger" lost two killed and had six wounded; the "Drake" lost forty-two killed and wounded. Against great odds John Paul Jones still remained victorious.

The people on shore heard the cannonading cease and saw the great clouds of gold-tinted smoke roll away to the south. There they saw the two vessels locked as if in an embrace of death and a great cheer went up.

They thought the "Drake" invincible. The gray of twilight turned to black, and the ships vanished like spectres in the darkness. But late that night some fishermen in a boat came ashore with a sail from the store-room of the "Drake." They said it had been given them by John Paul Jones. The people knew then that the "Drake" had been captured.

When the "Ranger" returned with her prizes to Brest, and his people told the tale of Paul Jones's victory, France was electrified. Neither in France nor in England would they at first believe it. France made him her hero. England offered ten thousand guineas for his head.

A STRUGGLE TO THE DEATH

NEVER, since the beginning of time, has there been a fiercer sea-fight than that between the "Bonhomme Richard" and the "Serapis." No struggle has been more dogged—no victory greater.

Three—four times during the night-long battle any other man than Paul Jones would have struck his colors. His main-deck battery and crews blown to pieces—his water-line gaping with wounds—his sides battered into one great chasm—still he fought on. His prisoners released—his masts tottering—his rudder gone—his ship afire below and aloft, his resistance was the more desperate. The thought of surrender never occurred to him.

After taking the "Drake" in a gallant fight, burning Whitehaven, and terrorizing the whole British coast, Paul Jones went to Paris, where a commission to the converted East Indiaman, the "Bonhomme Richard," awaited him. Putting her in the best shape possible, he boldly steered across for English waters. Paul Jones thirsted for larger game.

A STRUGGLE TO THE DEATH

When Captain Pearson, with the new frigate "Serapis," on a fine September afternoon in 1779, sighted Paul Jones, he signalled his merchant convoy to scatter, and piped all hands, who rushed jubilantly to quarters. The opportunity of his life had come, for the capture of the rebel frigate meant glory and a baronetcy. But he reckoned without his host.

Across the oily waters came the cheery pipes of the boatswain's mate of the "Richard" as Jones swung her up to meet her adversary, and Pearson knew his task would not be an easy one. The wind fell so light that the sun had sunk behind the light on Flamborough Head before the ships drifted up to fighting distance, and it was dark before they were ready to come to close quarters. On the "Bonhomme Richard," Jones's motley crew, stripped to the waist, were drawn up at the guns, peering out through the ports at the dark shadow on the starboard bow they were slowly overhauling.

The decks were sanded, the hammocks piled around the wheel, and there at the break of the poop stood the captain, trumpet in hand, turning now and then to give an order to Richard Dale or his midshipmen, quiet and composed, with the smile on his face men saw before the fight with the "Drake." The clumsy hulk rolled to the ground-swell, and the creaking of the masts and clamping of the sheet-blocks were all that broke the silence of the night. No excitement was apparent, and the stillness seemed the greater for an occasional laugh from the gunners, or the rattle of a cutlass newly settled in its sheath.

Then close aboard from out the blackness came a voice,—

"What ship is that?"

Paul Jones moved to the lee mizzen-shrouds and slowly replied,—

"I can't hear what you say."

He wanted all of his broadside to bear on the Englishman.

"What ship is that? Answer, or I shall fire."

The moment had arrived. For answer Jones leaned far over the rail of the poop and passed the word. A sheet of flame flashed from one of the "Richard's" after eighteen-pounders, followed by a terrific broadside which quaked the rotten timbers of the "Richard" from stem to stern. At the same time the guns of the "Serapis" were brought to bear, and her side seemed a mass of flame.

On the "Richard," two of the eighteen-pounders burst at this first broadside, killing their crews, heaving up the deck above, and driving the men from the upper tier. The others cracked and were useless. In this terrible situation Paul Jones knew the chances for victory were against him, for he had thought his lower battery his mainstay in a broadside fight.

But if he felt daunted his men did not know it, for, amid the hurricane of fire and roar of the guns, his ringing voice, forward, aft, everywhere, told them that victory was still theirs for the gaining. He ordered all of the men from the useless battery to the main deck ; and it was well he did so,—for so terrific was the fire that the six ports of the "Bonhomme Richard" were blown into one, and the shot passed clear through the ship, cutting away all but the supports of the deck above. No one but the marines guarding the powder-monkeys were left there, but they stood firm at their posts while the balls came whistling through and dropped into the sea beyond. But the fire of Paul Jones's battery did not slacken for a moment. There seemed to be two men to take the place of every man who was killed, and he swept the crowded deck of the "Serapis" from cathead to gallery.

In the meanwhile, the "Serapis," having the wind of the "Richard," drew ahead, and Pearson hauled his sheets to run across and rake

Jones's bows. But he miscalculated, and the American ran her boom over the stern of the Englishman. For a moment neither ship could fire at the other, and they hung together in silence, fast locked in a deadly embrace. Jones's crew, eager to renew the battle, glared forward at the shimmering battle-lanterns of the Englishman, cursing because their guns would not bear. The smoke lifted, and Paul Jones, who was deftly training one of his guns at the main-mast of the "Serapis," saw Pearson slowly climb up on the rail. The silence had deceived the Englishman, and his voice came clearly across the deck,—

"Have you struck?"

A harsh laugh broke from the "Richard."

"Struck!" Paul Jones's answer came in a roar that was heard from truck to keelson. "I haven't begun to fight yet!"

A cheer went up that drowned the rattle of the musketry from the tops, and the fight went on. Swinging around again the jib-boom of the "Serapis" came over the poop so that Paul Jones could touch it. Rushing to the mast, he seized a hawser, and quickly taking several turns with it, lashed the bowsprit of his enemy to his mizzen-rigging. Grappling-irons were dropped over on the enemy—and the battle became a battle to the death.

"Well done, lads; we've got her now." And Jones turned to his nine-pounders, which renewed their fire. Both crews fought with the fury of desperation. The men at the guns, stripped to the buff, grimed and blackened with powder, worked with extraordinary quickness. Every shot told. But the fire of the "Serapis" was deadly, and she soon silenced every gun but Jones's two nine-pounders, which he still worked with dogged perseverance. He sent Dale below to hurry up the powder charges. To his horror Dale found that the master-at-arms, knowing the ship to be sinking, had released a hundred English prisoners. The situation was terrifying. With foes within and without, there seemed no hope. But Dale, with ready wit, ordered the prisoners to the pumps and to fight the fire near the magazine, telling them that their only hope of life lay in that. And at it they went, until they dropped of sheer exhaustion.

The doctor passed Dale as he rushed upon deck. "Sir," said he to Jones, "the water is up to the lower deck, and we will sink with all hands in a few minutes."

Jones turned calmly to the doctor, as though surprised. "What, doctor," said he, "would you have me strike to a drop of water? Here, help me get this gun over."

The surgeon ran below, but Jones got the gun over, and served it, too.

To add to the horror of the situation, just at this moment a ball from a new enemy came screaming just over the head of Paul Jones, and the wind of it knocked off his hat. The carpenter, Stacy, ran up breathlessly.

"My God, she's firing on us—the 'Alliance,' sir!" And the captain glanced astern where the flashes marked the position of the crazy Landais, firing on his own consort.

If ever Paul Jones had an idea of hauling his colors, it must have been at this moment.

He had been struck on the head by a splinter, and the blood surged down over his shoulder—but he didn't know it.

Just then a fear-crazed wretch rushed past him, trying to find the signal-halyards, crying wildly as he ran,—

"Quarter! For God's sake, quarter! Our ship is sinking!"

Jones heard the words, and, turning quickly, he hurled an empty pistol at the man, which struck him squarely between the eyes, knocking him headlong down the hatch.

Pearson heard the cry. "Do you call for quarter?" he shouted.

For answer Paul Jones's nine-pounder cut away the rail on which he was standing.

"YARD-ARM TO YARD-ARM"

Then came the turn in the fight. Horrible as had been the slaughter on the "Richard," the quick flashes from his tops told Paul Jones that his marines had not been placed aloft in vain. He saw the crew on the spar-deck of his enemy fall one by one and men fleeing below for safety. Raising his trumpet, he cheered his topmen to further efforts. In their unceasing fire lay his only hope.

One of them in his maintop with great deliberateness laid aside his musket and picked up a leather bucket of hand grenades. Jones watched him anxiously as, steadying himself, he slowly lay out along the foot-rope of the main-yard. His captain knew what he meant to do. He reached the lift, which was directly over the main hatch of the "Serapis." There he coolly fastened his bucket to the sheet-block, and, taking careful aim, began dropping his grenades down the open hatchway. The second one fell on a row of exposed powder charges. The explosion that followed shook sea and sky, and the air was filled with blackened corpses. The smoke came up in a mighty cloud, and soon the forks of flame licked through it and up the rigging.

That was the supreme moment of Paul Jones's life, for he knew that victory was his.

The fire from the "Serapis" ceased as if by

magic. The explosion had blown a whole battery to eternity, and, as the smoke cleared a little, he could see the figure of Pearson leaning against the pin-rail, almost deserted, his few men running here and there, stricken mad with fear. Then the English captain stumbled heavily, as though blind, over the slippery deck towards the mizzen, where the flag had been nailed, and with his own hands tore it frantically from the mast.

A mighty victory for Paul Jones it was. But now, as the flames mounted higher through the rifts of smoke, he could see at what a cost. His dead lay piled upon the poop so that he could not get to the gangway. His masts were shot through and through, and strained at the stays at every lift of the bow. The fire, though beaten from the magazine, still burst from the forward hatches, firing the tangled rigging and outlining them in its lurid hues against the black beyond. The water had risen, and the freshening breeze lashed the purple foam in at the lower-deck ports. For hours the men fought against their new enemy; but towards five in the morning their captain decided that no human power could save her. He then began moving his wounded and prisoners to the "Serapis."

The first gray streaks of dawn saw Paul

Jones upon the poop of the "Serapis," looking to the leeward, where the "Richard" lay rolling heavily. Her flag, shot away again and again, had been replaced and floated proudly from its staff. Lower and lower she sank into the water, mortally wounded, a heavy swell washing in at the lower gun-ports. At length, heaving her stern high in the air, her pennant fluttering a last defiance to the captured "Serapis," she slowly disappeared, dying grandly as she had lived.

After Pearson's release, the British government offered ten thousand guineas for Paul Jones, dead or alive. Forty-two British frigates chased him and scoured the Channel; but Jones passed within sight of them, the American flag flying at the mast, and reached France in safety, where he became the hero of the hour. And so long as the Stars and Stripes fly over American war-ships will the men who know hold up as their ideal of a dogged warrior and gallant seaman the hero of Flamborough—Paul Jones.

THE TERRIER AND THE MASTIFF

THE first of the great American captains to give his life to the cause of liberty was Nicholas Biddle. And the action in which he lost it is the finest example of daring and hardihood in the little known pages of naval history. His part in that glorious action must ever remain unknown as to its details since but five out of his crew remained alive to tell of it, and we are chiefly indebted to the British accounts for the information which has been handed down.

Nicholas Biddle began his naval career by being shipwrecked on a desert shoal at the age of thirteen. But being rescued, with his four companions, at the end of two months, his ardor was so little dampened that as soon as opportunity offered he immediately went forth in search of further adventures on the sea. A war between England and Spain being imminent, he went to London, and succeeded in getting a midshipman's warrant on the ship of Captain—afterwards Admiral—Sterling.

But just before the declaration of independence of his own country, a voyage of discovery to the North Pole was proposed by the Royal Geographical Society, and this opportunity seemed to hold forth infinitely more possibilities for advancement than the daily port routine of a British frigate of war.

So, Admiral Sterling refusing Biddle's mild request to be transferred to one of the vessels, the young man took it upon himself to doff his gold-laced uniform and present himself upon the "Carcase" in very shabby sailor clothes, upon which he was forthwith entered upon her books as a sailor before the mast. He was in glorious company, though, for Horatio Nelson—afterwards to be the greatest admiral England has ever known—shared his humble lot as a jacky, although his prospects in the service were more brilliant than Biddle's. The expedition, having accomplished its purpose, returned to England in 1774, both young Nelson and Biddle having been appointed coxswains for meritorious service.

When hostilities in the United States began, Biddle, of course, resigned from the British navy and offered his services to the Continental Congress. His first commission was the command of the "Camden," a galley fitted out by the State of Pennsylvania for the defence of the Delaware River. He was then made a captain in the naval service, and took command of the "Andrew Doria," of fourteen guns and one hundred and thirty men.

Just before Commodore Hopkins's fleet hoisted anchor, Biddle had an opportunity to show his intrepidity in a very personal way. Two men who had deserted from his vessel had been taken and were placed in prison at Lewistown. Biddle sent an officer and a squad of men ashore to bring them off. But the officer returned to the ship and reported that the deserters had joined with the other prisoners, and barricaded the door, swearing that no man alive would take them. Biddle put on his side-arms and, taking only a young midshipman with him, went at once to the prison. The door was tightly barred from the inside, and the prisoners, led by one of the deserters named Green, shook their fists and pointed their weapons at him. Some of the more venturesome of the townsfolk, who only needed a resolute leader, now smashed down the door at the naval officer's directions, and Biddle, drawing both his pistols, quickly stepped within the opening. Green stood in front of his ill-favored companions, his eye gleaming villanously down the barrel of his flint-lock. Without moving his eye from the man, and planting himself squarely in the doorway, Biddle said, steadily,—

"Now, Green, if you don't take good aim, you are a dead man!"

There was a moment's pause, after which the pistol fell a little, and finally, under the resolute attitude of his captain, the fellow broke down. He was completely awed, and at Biddle's command dropped his pistol to the floor and allowed himself to be conducted to the ship. Their leader cowed, the remainder of the prisoners permitted the Lewistown militia, who had recovered from their fright, to come in and make them fast again.

This incident had its moral effect upon his men, and never again, when they learned to know him, was Biddle troubled with disaffection among his crew. The fury with which they went into the fights that followed showed how much he was a man after their own hearts.

After Commodore Esek Hopkins's unsuccessful encounter with the British fleet, the "Andrew Doria" put to sea and cruised off the coast of Newfoundland. Biddle captured a prize laden with arms and ammunition, which he carried to port, where they greatly strengthened Washington's army, which was badly in need of supplies of all kinds. He captured a transport and four hundred British soldiers, and made a great number of merchant prizes. He would have taken more, but he only had five men left aboard to take the "Doria" back to Philadelphia.

The Congress had authorized the building of several new frigates, and one of these, the "Randolph," of thirty-two guns, was just off the stocks. Biddle was made commander of her, and set immediately about finishing her and making her ready for sea. He had great difficulty in getting a crew, as privateering, where the prizes were greater and ship actions less frequent, proved more attractive to the adventurous spirits of the day. Congress, however, drafted a number of men from the army, and the crew was completed by the enlistment of volunteers from among the prisoners taken on prizes. After many difficulties with this motley crew, Biddle at last got to sea in February, 1777.

The men of his old crew were with him to a man, but many of the volunteers were shoal-water sailors, and his army recruits didn't know a sheet from a buntline. So when he ran into a Hatteras gale a few days out, the "Randolph" carried away her masts, and was altogether so uncomfortable a wreck that the volunteers mutinied, and Biddle had a hard time getting into Charleston harbor. He succeeded at last in refitting and in instilling some of the man-of-war spirit into his crew, sailing at last for the West Indies. Then his luck turned for the better, and he sighted the English ship "True Briton," twenty guns, convoying three merchantmen. Without accident he succeeded in taking them and in bringing all four prizes safe and sound into Charleston harbor. This was the first capture of the navy in the South, and, as the prizes were again liberally supplied with arms, the capture was doubly welcome. So much did Congress appreciate this affair that they had a medal struck off in Biddle's honor. The British hearing of this exploit of the "Randolph," sent a fleet south, and succeeded in blockading her at Charleston for a time.

The State of South Carolina got ready a fleet in the hope of raising the blockade, but before they could get to sea the Englishmen had disappeared.

In February, 1778, Biddle went out with a little fleet composed of the "General Moultrie," 18, the "Polly," 16, and the "Fair American," 14, in search of the British squadron. But missing them, they only succeeded in taking a few merchant vessels of the enemy. They boarded a number of Dutch and French ships, and Biddle knew that before long they must fall in with some of the enemy. To Captain Blake, who was dining with him, he said, "I would not be surprised if my old ship should be out after us. As to anything that carries her guns upon one deck, I think myself a match for her."

On the afternoon of the 7th of March, a sail was made out to windward, and they sailed up to examine her. As she came down with the wind she was made out to be square-rigged; but, bows on, she looked rather like a sloop than a frigate. A short time later she could be made out more plainly a man-of-war,—evidently of the enemy,—coming down speedily, and, from the way she was sailing, able to outfoot any of the squadron. Biddle could see that she stood well out of the water; but a small frigate might do that And if she was only a frigate of forty guns or under, he promised himself a great battle that day. But if she were a ship of the line, not only the "Randolph" but the smaller vessels were in great danger, for nothing save a craft somewhere near her size could resist the broadsides of the two heavy gun-tiers.

He quickly made his resolution. Signalling to the fleet of cruisers and prizes to go about, he himself took the deck and sent the little "Randolph" boldly down towards the stranger.

On she came, bowing majestically over the water, never making a sign until nearing gunshot distance, when the sound of the pipes and the calls on her deck showed that she was clearing ship for action. Biddle had been prepared for an hour. Now, as she came a little closer to the wind, the American captain discovered what he had suspected—two long lines of muzzles running out of her leeward ports.

She was a line-of-battle-ship, then.

He clinched his jaws and looked over his shoulder to where the prizes were scurrying away in the gathering darkness. They at least would be safe. But he did not shift his course a point, sailing on until the canvas of the great ship seemed to tower far above the little spars of his own vessel. The men of the "Randolph" were aghast at the action of their captain. To them an English "Sixty-Four" was the epitome of all that was powerful upon the seas. Biddle thought so, too; but there was nothing of timidity in his voice as he bade his gunners stand by to train upon her. He knew that this battle would be his last, for he resolved in those few moments that he would not give up his ship while one plank of her remained above water. The enemy might blow him out of the water and send him to the bottom, but before she did it he would give them such a lesson in patriotism that the world would not easily forget it.

His men guessed something of what was in his mind, and by the time the big ship hove close aboard they were keyed up to the fighting pitch, waiting with the utmost impatience for the first shot to be fired. The dusk had fallen, but the great loom of the sails of the English frigate showed plainly as she came closer. They were scarcely a pistol-shot apart when a figure on the Englishman mounted the hammock nettings aft, and a voice came clearly across the water,—

"Ahoy, the frigate!"

Biddle paused a moment to gain time, and then giving a word to his division officers, lifted his speaking-trumpet,—

"What ship is that?"

"His Britannic Majesty's ship-of-the-line 'Yarmouth,' Captain Vincent. Who are you? Answer, or I will be compelled to fire."

Another pause as Biddle directed the American colors to be run up to the mast, and then said,—

"This is the American Continental ship 'Randolph,' Captain Biddle!"

Without the pause of a second a tremendous broadside was poured into the Englishman, and in a moment the battle was on.

Biddle had gained a slight advantage in position by waiting as he did, and the "Randolph's" broadsides did great execution on the crowded decks of her adversary. But the "Yarmouth" men sprang to their guns, and in a few moments were firing their tremendous broadside of thirty guns as fast as they could be served and run out.

On the "Randolph" Biddle's men were working well, but the crashing of the shot and the flying splinters were terrific. In fifteen minutes the decks were covered with the bodies of dead and dying men, and the surgeon and his mate below in the cockpit, covered with blood, were laboring to help such of those as could be aided, and the decks, in spite of the sand, were so slippery that as the ship rolled it was difficult to stand upright upon them. Many of the guns of one of the broadsides were disabled, and there was not a gun that had a full crew to man it.

Biddle walked to and fro from one battery to another, lending a word here and a hand there, acting as sponger or tackle or handspikeman, wherever he was most needed. The men fought with the energy of despair—the despair of the dying. If they were to die, they would die hard, and the guns were loaded as though they would fire as many times as they could in the short time left them. The English aimed more deliberately. But when the dreaded broadside came, it dealt a blow that shook the smaller ship from stem to stern.

Biddle, although badly wounded, refused to leave the deck, and, ordering a stool to be placed where he could best direct the firing, sat calmly down, though in great agony, and gave the orders to his officers, who repeated them to the men.

* * * * * *

It has never been discovered just what happened on the "Randolph." In spite of her losses, she was keeping up her fire wonderfully, when, with scarcely a warning of any kind, she blew up.

The force of the explosion was so great that the ship split in two, and sank immediately. The air was filled with guns, spars, and the blackened bodies of men, many of which fell upon the deck of the "Yarmouth." An American ensign, neatly rolled in a ball, ready to be sent aloft on the "Randolph" if the others had been shot away, fell on the quarter-deck of the Englishman unsinged.

That national emblem was all, save a spar or two, that remained of the "Randolph." Captain Biddle and three hundred and ten of her crew of three hundred and fifteen were blown to pieces and drowned. Four days later the "Yarmouth," cruising near the same place, discovered a piece of the wreck to which five men, more dead than alive, had managed to cling.

The "Randolph" was lost, but the "Yarmouth" was so badly cut up that she could not follow the chase, and was obliged to lay to for repairs. What, if any, difference there might have been had the "Randolph" not been destroyed by explosion from within it is not easy to say; but all authorities agree that the fight, while it lasted, was one of the most determined in history. Captain Biddle at the time of his death was but twenty-eight years old, and the infant navy and the colonies lost one of their most intrepid officers and gallant seamen.

DECATUR AND THE "PHILADELPHIA"

IT was on the deck of the "Enterprise," before Tripoli, in 1804. The crew had been called aft, and Decatur, smiling, stood on his quarter-deck.

"My men," said he, "the 'Philadelphia' is in the hands of the enemy. A few days from now and we may see American guns turned against American sailors. The commodore has given us permission to sail in and blow her up. Will you go?"

Into the air flew a hundred caps, and three wild American cheers were the answer

"I can't take you all," he explained; "the expedition is a dangerous one. We are going under the broadsides of the enemy, and I only want those of you who are ready. Now, lads, any of you who are willing to go, take one step aft."

Without a second's pause the crew of the "Enterprise," to a man, stepped out; then, fearful lest others should get in the front rank, came towards the young commander in a body, elbowing and swearing at one another lustily.

Decatur smiled. With such a spirit there was nothing he might not accomplish. He picked out sixty-two of his youngest and steadiest men, each of them touching his tarry cap with a grateful "Thank'ee, sir," as Decatur called his name.

That afternoon they tumbled joyfully down into a captured ketch, which had been named the "Intrepid," and, stores aboard, hoisted their three-cornered sail for the harbor of Tripoli. As they hauled off, Decatur went below to see that all his supplies and combustibles were stored, when Midshipman Lawrence came towards him somewhere from the depths of the fore-hold, pushing along by the scruff of the neck a youngster, who was crying bitterly.

"I found this stowaway, sir," said Lawrence, with a smile.

"Please, sir," sobbed the boy, "don't send me back. I want to see this 'ere fight, and I ain't going to do no harm. Don't send me back, sir."

Decatur had looked up with a fierce frown, but the anxiety on the lad's face was pathetic, and he smiled in spite of himself.

"You can go," he laughed, "but I'll put you in the brig—when we get back."

On that six days' voyage to Tripoli the wind

blew a hurricane, and the masquerade of the American tars seemed likely to end in disaster, without even a fight for their pains. But as they sighted the coast the sea went down, and the arrangements were completed. The yellow sails of the "Siren," their consort, hove again into sight, and by the afternoon of the 16th of February the two vessels were bearing down upon the dark line that lay shimmering purple under the haze of the southern sky.

The sun dropped down, a ball of fire, into the western sea, and by eight o'clock the towers of the bashaw's castle loomed dark against the amber of the moonlit sky. To the left the stately spars of the doomed frigate towered above the rigging in the harbor, and floating at her truck was the hated insignia of the enemy.

The piping northern breeze bellied the crazy sail of the ketch and sent the green seas swashing under the high stern, speeding them good luck on their hazardous venture. Catalano, the pilot, stood at the helm, swinging the clumsy tiller to meet her as she swayed. By his side was a tall figure, a white burnoose about his shoulders and a fez set jauntily on his head—Decatur. Four others, in unspeakable Tripolitan costumes, lounged about the deck or squatted cross-legged. But the delusion went no further. For one of them, Reuben James, was puffing at a stubby black pipe, and another spat vigorously to leeward. The others were below, lying along the sides, sharpening their cutlasses.

On they sped, Catalano heading her straight for the frigate. As the harbor narrowed and the black forts came nearer, they could see the dusky outlines of the sentries and the black muzzles that frowned on them from the battlements. Over towards the east faint glimmers showed where the town was, but the wind had now fallen low, and the lapping of the water along the sides alone awoke the silence. A single light shone from the forecastle of the frigate, where the anchor watch kept its quiet vigil. She swung at a long cable, a proud prisoner amid the score of watchful sentinels that encircled her.

As placid as the scene about him, Decatur turned to the pilot and gave a low order. The helm was shifted and the tiny vessel pointed for the bowsprit of the "Philadelphia." Nearer and nearer they came, until scarcely a cable's length separated them. They saw several turbaned heads, and an officer leaned over the rail, puffing lazily at a cigarette. He leisurely took the cigarette from his mouth, and his voice came across the quiet water of the harbor,—

"Where do you come from?" he hailed.

Catalano, the pilot, answered him in the lingua Franca of the East,—

"The ketch 'Stella,' from Malta. We lost our anchors and cables in the gale, and would like to lie by during the night."

The Tripolitan took another puff, and an ominous stir, quickly silenced, was heard down in the hold of the ketch. It seemed an eternity before the answer came,—

"Your request is unusual, but I will grant it," said the Tripolitan, at last. "What ship is that in the offing?"

The officer had seen the "Siren," which hovered outside the entrance of the harbor.

"The British ship 'Transfer,'" said Catalano, promptly.

The ketch was slowly drifting down until a grappling-iron could almost be thrown aboard. Right under the broadside she went, and a line of dark heads peered over the rail at her as she gradually approached the bow.

The chains of the frigate were now almost in the grasp of Reuben James, on the forecastle, when the wind failed and a cat's-paw caught the ketch aback. Down she drifted towards the terrible broadside. But at a sign from Decatur the eager Lawrence and James got into a small boat and carried a line to a ring-bolt at the frigate's bow. A boat put out from the "Philadelphia" at the same time. But Lawrence coolly took the hawser from the Tripolitan—"to save the gentleman trouble," he explained—and brought it aboard the "Intrepid." A moment more, and the ketch was warping down under the "Philadelphia's" quarter. It was a moment of dire peril. The slightest suspicion, and they would be blown to pieces.

Decatur leaned lightly against the rail, but his hand grasped his cutlass under his robe so that the blood tingled in his nails and his muscles were drawn and tense. Morris and Joseph Bainbridge stood at the rigging beside him, trembling like greyhounds in leash.

Suddenly they swung around and shot out from under the shadow into a yellow patch of moonlight. The watchful eyes above the rail saw the anchor and cables and the white jackets of the sailors below decks as they strove to hide themselves in the shadows. One glance was enough. In an instant the ship resounded with the thrilling cry, "Americano! Americano!"

At the same moment the "Intrepid" ground up against the side of the frigate. In an instant, as if by magic, she was alive with men. Throwing off his disguise, and with a loud cry of "Boarders, away!" Decatur sprang for the mizzen-chains. And now the hot blood of fighting leaped to their brains. The long agony of suspense was over. Lawrence and Laws sprang for the chain-plates and hauled themselves up. Decatur's foot slipped, and Morris was the first on deck. Laws dashed at a port, pistols in hand. Nothing could withstand the fury of the charge, and over the rail they swarmed, cutlasses in teeth, jumping over the nettings, and down on the heads of the Tripolitans below. Though Morris was first on deck, Decatur lunged in ahead of him, bringing down the Tripolitan officer before he could draw his sword. One of them aimed a pike at him, but he parried it deftly, and Morris cut the fellow down with a blow that laid his shoulder open from collar to elbow.

Though surprised, the Tripolitans fought fiercely. They had won their title of "the best hand-to-hand fighters in the world" in many a hard pirate battle in the Mediterranean. Around the masts they rallied, scimetars in hand, until they were cut or borne down by the fury of their opponents.

After the first order, not a word was spoken and not a shot was fired. The Americans needed no orders. Over the quarter-deck they swept—irresistible, clearing it in a trice.

DECATUR BOARDS THE "PHILADELPHIA"

Overwhelmed by the fierce onslaught, the Tripolitans fled for life, the sailors driving them up on the forecastle and overboard in a mass, where their falling bodies sounded like the splash of a ricochet.

So swift was the work that in ten minutes no Tripolitans were left on the deck of the frigate but the dead. Not a sailor had been killed. One man had been slashed across the forehead, but he grinned through the blood and fought the more fiercely. Then the watchers out on the "Siren" saw a single rocket go high in the air, which was Decatur's signal that the "Philadelphia" was again an American vessel.

In the meanwhile the combustibles were handed up from the ketch with incredible swiftness, and the work of destruction began. Midshipman Morris and his crew had fought their way below to the cock-pit and had set a fire there. But so swiftly did those above accomplish their work that he and his men barely had time to escape. On reaching the upper deck, Decatur found the flames pouring from the port-holes on both sides and flaring up red and hungry to seize the tar-soaked shrouds. He gave the order to abandon, and over the sides they tumbled as quickly as they had come. Decatur was the last to leave the deck. All the men were over, and the ketch was drifting clear, while around him the flames were pouring, their hot breath overpowering him. But he made a jump for it and landed safely, amid the cheers of his men.

Then the great oars were got out, eight on a side, and pulling them as only American sailor-men could or can, they swept out towards the "Siren."

The Tripolitans ashore and on the gunboats had hastened to their guns, and now, as the ketch was plainly seen, their batteries belched forth a terrific storm of shot that flew across the water. The men bent their backs splendidly to their work, jeering the while at the enemy as the balls whistled by their heads or sent the foam splashing over them. Out they went across the great crimson glare of the fire. It was magnificent. The flames swept up the shrouds with a roar, catching the woodwork of the tops and eating them as though they were tinder. She was ablaze from water to truck, and all the heavens were alight,—aglow at the splendid sacrifice. Then to the added roar of the batteries ashore came the response from the guns of the flaming ship, which, heated by the fierce flames, began to discharge themselves. But not all of them were fired so, for in a second all eyes were

THE DANGER OF THE "INTREPID"

dazzled by a blazing light, and they saw the great hull suddenly burst open, with huge streaks of flame spurting from between the parting timbers. Then came a roar that made the earth and sea shudder. The fire had reached the magazine.

The waves of it came out to the gallant crew, who, pausing in their work, gave one last proof of their contempt of danger. Rising to their feet, they gave three great American cheers that echoed back to the forts while their guns thundered fruitlessly on.

Decatur and his men were safe under the "Siren's" guns.

Is it any wonder that Congress gave Decatur a sword and made him a captain, or that Lord Nelson called this feat "the most daring act of any age"?

THE BIGGEST *LITTLE* FIGHT IN NAVAL HISTORY

IT should have been renown enough for one man to have performed what Nelson was pleased to call "the most daring act of any age." But the capture of the "Philadelphia" only whetted Decatur's appetite for further encounters. He was impetuous, bold even to rashness, and so dashing that to his men he was irresistible. But behind it all—a thing rare in a man of his peculiar calibre—there was the ability to consider judiciously and to plan carefully as well as daringly to execute. His fierce temper led him into many difficulties, but there was no cruelty behind it; and the men who served with him, while they feared him, would have followed him into the jaws of death, for they loved him as they loved no other officer in the American service. Once while the frigate "Essex," Captain Bainbridge, lay in the harbor at Barcelona, the officers of the American vessel suffered many petty indignities at the instance of the officers of the Spanish guardship. Having himself been subjected to a slight from the

Spanish commander, Lieutenant Decatur took the bull by the horns. He bade his coxswain pull to the gangway of the Spaniard, and he went boldly aboard. His lips were set, for he had resolved upon his own responsibility to make an immediate precedent which would serve for all time. The Spanish commander, most fortunately, was absent. But Decatur none the less strode aft past the sentry to the gangway and, lifting his great voice so that it resounded from truck to keelson, he shouted,—

"Tell your comandante that Lieutenant Stephen Decatur, of the 'Essex,' declares him to be a scoundrelly coward, and if Lieutenant Decatur meets him ashore he will cut his ears off."

So among the men of the squadron Decatur came to be known as a man who brooked nothing and dared everything.

But when the crusty Preble took command in the Mediterranean he was not over-impressed with the under-officers of his command. Not one of the lieutenants was over twenty-four and none of those higher in authority had turned thirty. Decatur and Somers were twenty-five; Charles Stewart was only twenty-six, and Bainbridge the younger; Morris and Macdonough were barely out of their teens.

It was not the custom of the commander-in-chief to mince his words. So sparing himself the delicacy of secluding himself behind the saving bulkheads of the after-cabin he swore right roundly at his home government for sending him what he was pleased to call "a parcel of d— school-boys." He was a martinet of the old style, and believed in the school of the fo'c's'le, and not in young gentlemen whose friends at home sent them in by the ports of the after-cabin. He held the youngsters aloof, and not until he had tried them in every conceivable fashion would he consider them in his councils. A year had passed, and Decatur, Morris, Bainbridge, Macdonough, and Somers had helped to add glorious pages to naval history, before the old man, with a smile to Colonel Lear, the consul, consented to say,—

"Well, after all, colonel, they are very good school-boys!"

Although Decatur's success in the destruction of the "Philadelphia" had removed a dangerous auxiliary battery from the harbor of Tripoli, the bashaw was far from overawed, and, with the officers and crew of the "Philadelphia" as hostages, declined to consider any terms offered by the Americans; and so it was resolved by Commodore Preble to make an attempt upon the Tripolitan batteries and fleet. The Americans had the "Constitution,"—"Old Ironsides,"—Commodore Preble, and six brigs and schooners mounting twelve and sixteen guns each. Preble had also succeeded in borrowing from "the most gracious king of the Sicilies," who was then at war with the bashaw, two bomb-vessels and six single gunboats,—quite a formidable little force of a hundred and thirty-four guns and about a thousand men.

It was not until the morning of the 3d of August, 1804, that the weather, which had been very stormy, moderated sufficiently to allow the squadron to approach the African coast. The gunboats were unwieldy craft, flat-bottomed, and, as the sea made clean breeches over them, they were a dozen times in danger of sinking. But by ten o'clock the sky to the southward had lightened, and the heavy storm-clouds were blowing away overhead to the westward. "Old Ironsides" shook the reefs out of her topsails and, spreading her top-gallant-sails, she beat up for the entrance of the harbor of Tripoli with two of the gunboats in tow. Her tall spars, seeming almost to pierce the low-rolling clouds, towered far above the little sticks of the "Siren" and "Nautilus," which bore down directly in her wake. The sea had lashed out its fury, and, before the little fleet had reached the reef, the gray had turned to green, and here and there a line of amber showed where the mid-day sun was stealing through.

Stephen Decatur, on gunboat No. 4, had been given command of the left division of three gunboats. Casting off the tow-lines from his larger consorts, he got under weigh, and bore down for a rift between the reefs at the eastern entrance to the harbor, where the Tripolitan fleet, cleared for action, lay awaiting him. The wind was on his bow, and he was obliged to hold a course close to the wind in order to weather the point.

The gunboat lumbered uncertainly in the cross-sea, for she had no longer the steady drag of the "Constitution's" hawser to steady her. The seas came up under her flat bottom, and seemed to toss rather than swing her into the hollows. She was at best an unsteady gun-platform, and nice sail-trimming was an impossibility. But they got out their sweeps, and that steadied her somewhat. Great volumes of spray flew over the weather-bow as she soused her blunt nose into it, and the fair breeze sent it shimmering down to leeward.

Decatur stood aft by the helmsman, watching the quivering leeches, and keeping her well up into the wind. Beside him stood his midshipmen, Thomas Macdonough—afterwards to win a great victory of his own—and Joseph Thorn. Both of them had smelt powder before, and Macdonough had been one of the first on the deck of the ill-fated "Philadelphia." This was to be a different sort of a fight from any they had seen. It was to be man to man, where good play of cutlass and pike and youth and American grit might mean victory. Defeat meant annihilation. But youth is good at a game of life and death, and as they looked at Decatur there was never a moment's fear of the result. They leaned against the rail to leeward, looking past the foam boiling on the point to the spars of the African gunboats, and their eyes were alight with eagerness for battle.

The men were bending steadily to their sweeps. Most of them were stripped to the waist, and Decatur looked along the line of sinewy arms and chests with a glow of pride and confidence. There was no wavering anywhere in the row of glistening faces. But they all knew the kind of pirates they were going to meet,—reckless, treacherous devils, who loved blood as they loved Allah,—the best hand-to-hand fighters in the Mediterranean.

The ring of the cutlasses, loose-settled in their hangers, against the butts of the boarding-pistols was clear above the sound of the row-locks and the rush of the waters, while forward the catch of a song went up, and they bent to their work the more merrily.

As they came under the lee of the Tripolitan shore and the sea went down, Decatur ordered the long iron six-pounder cast loose. They had provided solid shot for long range at the batteries, and these were now brought up and put conveniently on the fo'c's'le. But for the attack upon the vessels of the fleet they loaded first with a bag of a thousand musket-balls. At point-blank range Decatur judged that this would do tremendous execution among the close-ranked mass of Tripolitans on the foreign vessels. His idea was not to respond to the fire of the enemy, which would soon begin, until close aboard, and then to go over the rail before they could recover from their confusion. He felt that if they did not make a wreck of him and batter up his sweeps he could get alongside. And once alongside, he knew that his men would give a good account of themselves.

But as they came up towards the point the wind shifted, and the head of the gunboat payed off. Even with their work at the sweeps, he now knew that it would be no easy matter for all the Americans to weather the point, for two of them were well down to leeward. But his brother, James Decatur, in gunboat No. 2, and Sailing-Master John Trippe, in gunboat No. 6, had kept well up to windward, and so he felt that he should be able to count on at least these two. As they reached the line of breakers, one of the gunboats to leeward, under Richard Somers, was obliged to go about, and in a moment the two others followed. Then the young commanders of the windward gunboats knew that if the attack was to be made they alone would have the glory of the first onslaught.

What Decatur feared most was that Preble, on the "Constitution," would see how terribly they were overmatched and signal the recall. But as they reached the point, Decatur resolutely turned his back to the flagship, and, putting his helm up, set her nose boldly into the swash of the entrance and headed for the gray line of vessels, three times his number, which hauled up their anchors and came down, gallantly enough, to meet him.

There was very little sound upon the gunboat now. The wind being favorable, the Americans shipped their sweeps, and sat watching the largest of the Tripolitan vessels, which was bearing down upon them rapidly. They saw a puff of white smoke from her fo'c's'le, and heard the whistle of a shot, which, passing wide, ricochetted just abeam and buried itself beyond. Thorn stood forward, waiting for the order to fire his long gun. But Decatur gave no sign. He stood watching the lift of the foresail, carefully noting the distance between the two vessels. Trippe and James Decatur had each picked out an adversary, and were bearing down as silently as he, in spite of the cannonade which now came from both the vessels and batteries of the Turks. The shots were splashing all around him, but nothing had been carried away, and the American jackies jeered cheerfully at the wretched marksmanship. As the Tripolitans came nearer, the Americans could see the black mass of men along the rails and catch the glimmer of the yataghans. Then Decatur ordered his own men to seize their pikes and draw their pistols and cutlasses.

At the word from Decatur, Thorn began training the fo'c's'le gun, which in the steadier sea would have a deadly effect. The distance was a matter of yards now, and a shot came ploughing alongside that threw spray all along the rail and nearly doused the match of the gunner of the fo'c's'le. But not until he could see the whites of the eyes of his adversaries did Decatur give the order to fire. As the big gun was discharged point-blank into the thick of the crowded figures, Decatur shifted his helm quickly and lay aboard the Tripolitan. So tremendous had been the execution of the musket-balls, and so quickly had the manœuvre been executed, that almost before the Tripolitans were aware of it the Americans were upon them. The few shots from the Turkish small arms had gone wild, but a fierce struggle ensued before the Americans reached the deck. At last Decatur, followed by Thorn, Macdonough, and twenty-two seamen, gained the fo'c's'le in a body, and the Tripolitans retreated aft.

The Tripolitan boat was divided amidships by an open hatchway, and for a moment the opposing forces stopped to catch their breath, glaring at one another across the opening. Decatur did not pause long. Giving them a volley of pistol-bullets at close range, he dashed furiously down one gangway, while Macdonough and Thorn went down the other, and, with a cheer, cut down the remaining Turks or drove them overboard. A half-dozen went down a forward hatch, and these were made prisoners.

It was a short fight, with an inconsiderable

loss to Decatur, but the Tripolitan dead were strewn all over the decks, and the Turkish captain was pierced by fourteen bullets. The Tripolitan flag was hauled down, and, taking his prize in tow, Decatur put his men at the sweeps again, to move farther out of the reach of the batteries.

By this time James Decatur and John Trippe had got into the thick of it. Following Stephen Decatur's example, they dashed boldly at the larger of the bashaw's vessels, and, reserving their fire for close range, they lay two of them aboard. John Trippe, Midshipman Henley, and nine seamen had gained the deck of their adversary, when the vessels drifted apart, and they were left alone on the deck of the enemy. But Trippe was the man for the emergency. So rapidly did they charge the Turks that their very audacity gave them the advantage, and Trippe finally succeeded in killing the Tripolitan commander by running him through with a boarding-pike. They fought with the energy of despair, and, although wounded and bleeding from a dozen sabre-cuts, struggled on until their gunboat got alongside and they were rescued by their comrades.

But the story of the treachery of the Turkish captain and Stephen Decatur's revenge for the death of his brother makes even the wonderful defensive battle of Trippe seem small by comparison.

James Decatur, having got well up with one of the largest of the Tripolitan vessels, delivered so quick and telling a fire with his long gun and musketry that the enemy immediately struck his colors. He hauled alongside and clambered up and over the side of the gunboat to take possession of her personally. As his head came up above the rail his men saw the Turkish commander rush forward and aim his boarding-pistol at the defenceless American. The bullet struck him fairly in the forehead, and Decatur, with barely a sound, sank back into his boat.

In their horror at the treachery of the Tripolitan, the Americans allowed the boat to sheer off, and the Turk, getting out his sweeps, was soon speeding away toward the protection of the batteries.

Stephen Decatur, towing his prize to safety, had noted the gallant attack, and had seen the striking of the Turkish colors. But not until an American boat darted alongside of him did he hear the news of the treacherous manner of his brother's death. The shock of the information for the moment appalled him, but in the place of his grief there arose so fierce a rage at the dastardly act that for a moment he was stricken dumb and senseless. His men sprang quickly when at last he thundered out his orders. Deftly casting off the tow-line of the prize, they hoisted all sail and jumped to their sweeps as though their lives depended on it. Macdonough's gun-crew were loading with solid shot this time, and, as soon as they got the range, a ball went screaming down towards the fleeing Tripolitan. The men at the sweeps needed little encouragement. They had heard the news, and they loved James Decatur as they worshipped his brother, who stood aft, his lips compressed, anxiously watching the chase. The water boiled under the oar-blades as the clumsy hulk seemed to spring from one wave-crest to another. Again the long gun spoke, and the canister struck the water all about the Turkish vessel. The Tripolitans seemed disorganized, for their oars no longer moved together and the blades were splashing wildly. Another solid shot went flying, and Decatur smiled as he saw the spray fly up under the enemy's counter. There would be no mercy for the Tripolitans that day. Nearer and nearer they came, until the Turks, seeing that further attemps at flight were useless, dropped their sweeps and prepared to receive the Americans. They shifted their helm so that their gun could bear, and the shot that followed tore a great rent in Decatur's foresail. But the Americans heeded it little more than if it had been a puff of wind, and pausing only to deliver another deadly discharge of the musket-balls at point-blank range, Decatur swung in alongside under cover of the smoke.

As the vessels grated together, Decatur jumped for the Tripolitan rigging, and, followed by his men, quickly gained the deck. Two Turks rushed at Decatur, aiming vicious blows with their scimetars; but he parried them skilfully with his pike, looking around him fiercely the while for the captain. As he thought of his brother dying, or dead, he swore that no American should engage the Turkish commander but himself. He had not long to wait. They espied each other at about the same moment, and brushing the intervening weapons aside, dashed upon each other furiously.

Decatur was tall, and as active as a cat. His muscles were like steel, and his rage seemed to give him the strength of a dozen. But the Mussulman was a giant, the biggest man in the Tripolitan fleet, and a very demon in power and viciousness. So strong was he, that as Decatur lunged at him with his boarding-pike he succeeded in wrenching it from the hand of the American, and so wonderfully quick that Decatur had hardly time to raise his cutlass to parry the return. He barely caught it; but in doing so his weapon broke off short at the hilt. The next lunge he partially warded by stepping to one side; but the pike of the Mussulman in passing cut an ugly wound in his arm and chest. Entirely defenceless, he now knew that his only chance was at close quarters, so he sprang in below the guard of the Turk and seized him around the waist, hoping to trip and stun him. But the Tripolitan tore the arms away as though he had been a stripling, and, seizing him by the throat, bore him by sheer weight to the deck, trying the while to draw a yataghan. The American crew, seeing things going badly with their young captain, fought in furiously, and in a moment the mass of Americans and Tripolitans were fighting in one desperate, struggling, smothering heap, above the prostrate bodies of their captains, neither of whom could succeed in drawing a weapon. The Turk was the first to get his dagger loose, but the American's death-like grasp held his wrist like a vise, and kept him from striking the blow. Decatur saw another Turk just beside him raise his yataghan high above his head, and he felt that he was lost. But at this moment a sailor, named Reuben James, who loved Decatur as though he were a brother, closed in quickly and caught on his own head the blow intended for Decatur. Both his arms had been disabled, but he asked nothing better than to lay down his life for his captain.

In the meanwhile, without relinquishing his grip upon the Turk, Decatur succeeded in drawing a pistol from the breast of his shirt, and, pressing the muzzle near the heart of the Tripolitan, fired. As the muscles of his adversary relaxed, the American managed to get upon one knee, and so to his feet, stunned and bleeding, but still unsubdued. The Tripolitans, disheartened by the loss of their leader, broke ground before the force of the next attack and fled overboard or were cut down where they stood.

The death of James Decatur was avenged.

The other Tripolitan gunboats had scurried back to safety, so Decatur, with his two prizes, made his way out towards the flagship unmolested. His victory had cost him dearly. There was not a man who had not two or three wounds from the scimetars, and some of them had cuts all over the body. The decks were like a slaughter-pen and the scuppers were running blood. But the bodies of the Tripolitans were ruthlessly cast overboard to the sharks; and by the time the Americans had reached the "Constitution" the decks had been scrubbed down and the wounded bandaged and roughly cared for by those of their comrades who had fared less badly.

Decatur, by virtue of his exploit in destroying the "Philadelphia," already a post-captain at the age of twenty-five, could expect no further immediate honors at the hands of the government; but then, as ever afterwards, he craved nothing but a stanch ship and a gallant crew. The service he could do his country was its own reward.

A DOUBLE ENCOUNTER

THE old "Constitution" was out on the broad ocean again! And when the news went forth that she had succeeded for the seventh time in running the blockade of the British squadrons, deep was the chagrin of the Admiralty. This Yankee frigate, still stanch and undefeated, had again and again proved herself superior to everything afloat that was British; had shown her heels, under Hull's masterly seamanship, to a whole squadron during a chase that lasted three days; and had under Hull, and then under Bainbridge, whipped both the "Guerriere" and the "Java," two of their tidiest frigates, in an incredibly short time, with a trifling loss both in men and rigging. She was invincible; and the title which she had gained before Tripoli, under Commodore Preble, when the Mussulman shot had hailed against her oaken timbers and dropped harmlessly into the sea alongside, seemed more than ever to befit her. "Old Ironsides" was abroad again, overhauled from royal to locker, with a crew of picked seamen and a captain who had the confidence of the navy and the nation.

Her hull had been made new, her canvas had come direct from the sail-lofts at Boston, and her spars were the stanchest that the American forests could afford. She carried thirty-one long 24-pounders and twenty short 32-pounders,—fifty-one guns in all, throwing six hundred and forty-four pounds of actual weight of metal to a broadside. Her officers knew her sailing qualities, and she was ballasted to a nicety, bowling along in a top-gallant-stu'n-sail breeze at twelve knots an hour.

The long list of her victories over their old-time foe had given her men a confidence in the ship and themselves that attained almost the measure of a faith; and, had the occasion presented itself, they would have been as willing to match broadsides with a British seventy-four as with a frigate of equal metal with themselves. They were a fine, hearty lot, these jack-tars; and, as "Old Ironsides" left the green seas behind and ploughed her bluff nose boldly through the darker surges of the broad Atlantic, they vowed that the frigate's last action would not be her least. The "Constitution" would not be dreaded by the British in vain.

For dreaded she was among the officers of the British North Atlantic squadron. As soon as it was discovered by the British Admiralty that she had passed the blockade, instructions were at once given out and passed from ship to ship to the end that every vessel of whatever class which spoke another on the high seas should report whether or not she had seen a vessel which looked like the "Constitution." By means of this ocean telegraphy they hoped to discover the course and intention of the great American, and finally to succeed in bringing her into action with a British fleet. By this time they had learned their lesson. Single frigates were given orders to avoid an encounter, while other frigates were directed to hunt for her in pairs!

Charles Stewart had been one of old Preble's "school-boy captains" before Tripoli, the second in command. He had been one to suggest the expedition to cut out or destroy the "Philadelphia," the envied command of which fell to Decatur. But he won distinction enough before the batteries there, and afterwards when he captured the French "Experiment," of a much heavier force and armament than his own, in a brilliant little action. He had entered the merchant service at thirteen, had been captain of a ship in the India trade at nineteen, and thus from his boyhood had been schooled in the finer points of rough-and-ready seamanship.

He was born in Philadelphia, in 1778, at a time when the blood of patriotism ran hot in the veins of the mothers as well as of the fathers of the race, and he then imbibed the principles he afterwards stood for so valiantly on sea and on land. On the frigate "United States," that "nursery of heroes," he had for messmates Stephen Decatur and Richard Somers; and Edward Preble gave him ideas of discipline that later stood him in good stead. He was, like Decatur, of an impetuous disposition; but he had learned what quick obedience meant to the service, and among the men on the "Constitution" it was known that infractions of duty would be quickly punished. The men tumbled quickly to the gear and handled the guns so smartly that with his picked seamen Stewart had not been out of sight of land a week before they attained a proficiency in manœuvre rarely surpassed on a man-of-war. It is related that once, having received an order from a superior officer to sail with his ship immediately, Stewart got under weigh, towing behind him his mainmast, which he had not had the opportunity to step.

Stewart was, of course, aware of the orders

which had been issued by the Admiralty, but with his ship in fine condition and provisioned for a long cruise he feared nothing that floated, whether one ship or two. In fact, just before leaving his young wife in Boston he had asked her what he should bring her home.

"A British frigate," said she, patriotically.

"I will bring you two of them," he said, smiling.

Stewart sailed to the southward, in the hope of falling in with some vessels in the India trade. For two months, in spite of their fitness, the men were daily exercised in all weathers at evolutions with the sails and great guns, and part of the day was given to cutlass-work and pistol-practice. No emergency drill was overlooked, and from reefing topsails to sending up spare spars or setting stu'n-sails they moved like the co-ordinated parts of a great machine. But one prize having been taken, however, Stewart set his course for the coast of Europe, to seek the lion, like Paul Jones, on his own cruising ground.

On February 18, 1815, just two months after leaving Boston, the "Constitution," being then near the Portuguese coast, sighted a large sail, and immediately squared away in pursuit. But hardly were they set on their new course before another sail hove up to leeward, and

Stewart quickly made down for her. Overhauling her shortly, she was discovered to be the British merchant ship "Susan," which he seized as a prize and sent back to Boston. Meanwhile the other sail, which afterwards proved to be the "Elizabeth," 74, had disappeared.

The following day the "Constitution" was holding a course to the southward from the coast of Spain toward Madeira. A group of her officers stood upon her quarter-deck, watching the scud flying to leeward. They were rather a discontented lot. They had been to sea two months, and beyond a few merchant prizes they had nothing to show for their cruise. It was not like the luck of "Old Ironsides." What they craved was action to put a confirmatory test to the metal they were so sure of. The fo'c's'le was grumbling, too; and the men who had been in her when she fought the "Guerriere" and the "Java" could no longer in safety boast of the glory of those combats.

Had they but known it, the "Elizabeth," 74, and the "Tiber," 38, in command of Captain Dacres, who had lost the "Guerriere," were but a few hours astern of them; and the "Leander," 50, the "Newcastle," 50, and the "Acasta," 40, whom they had so skilfully eluded at Boston, were dashing along from the westward in pursuit. The seas to the eastward, too, were swarming with other frigates (in couples), who were seeking her no less anxiously than she was seeking them.

Stewart was not so easily disheartened as his officers. He knew that the "Constitution" was in the very midst of the ships of the enemy. Had he not known it he would not have been there. He came on deck during the afternoon in a high good humor. He was a believer in presentiments, and said, jovially,—

"The luck of the 'Constitution' isn't going to fail her this time, gentlemen. I assure you that before the sun rises and sets again you will be engaged in battle with the enemy, and it will not be with a single ship."

The morning of the next day dawned thick and cloudy. Though well to the southward, the air was cold and damp. The wind was blowing sharply from the northeast, and the choppy seas sent their gray crests pettishly or angrily upward, where they split into foam and were carried down to mingle with the blur of the fog to leeward. Occasionally, in the wind-squalls, the rain pattered like hail against the bellying canvas and ran down into the lee-clews, where it was caught as it fell and whipped out into the sea beyond.

Two or three officers paced the quarter-deck, looking now and then aloft or to windward to see if the weather were clearing. Saving these, the fellows at the wheel, and the watch on deck, all hands were below on the gun-deck, polishing their arms or loitering in the warmth near the galley, where the cooks were preparing the mid-day meal.

During the morning watch, Stewart, for some reason which he was unable to give, save an unaccountable impulse, changed the course and sent the ship down sixty miles to the southwest. Shortly after noon the fog fell lower, and so thinned out at the mast-head that the lookout on the topsail-yard could soon see along its upper surface. At about one o'clock the welcome sound of "Sail, ho!" came echoing down through the open hatchways. While ordinarily the sighting of a sail so near the coast has no great significance, Stewart's prediction of a battle had aroused the men to a fever of impatience; and when they knew that a large sail, apparently a frigate, had been raised and that the fog was lifting, the watch below dropped their kits and tools and tumbled up on deck to have a glimpse of the stranger. Here and there wider rifts appeared in the fog-banks, and the midshipman of the watch, who climbed with a glass into the foretop, soon made her out to be a frigate bearing about two points on the port-bow.

Stewart came up from below and immediately crowded on top-gallant-sails and royals in pursuit. Before long the weather had cleared, so that they could make out the horizon to windward, and from the deck could dimly discern the hazy mass of the chase as she hung on the lee-bow, apparently motionless. In less than an hour the man at the mast-head reported another sail ahead of the first one, and noted that signals were being exchanged between them.

It was now almost a certainty that the vessels were those of the enemy. Forward the men were slapping one another on the back, and rough jokes and laughter resounded from the gun-deck, where the boys and stewards were clearing away the mess-dishes and stowing away all gear, in preparation for a possible action. On the quarter-deck wagers were freely offered on the character of the vessels, which looked to be frigates of 50 and 38. Stewart glanced aloft at the straining spars and smiled confidently.

By this time the nearer frigate bore down within the range of the glasses, and they could see that she was painted with double yellow lines, and apparently cut for fifty guns. As it afterwards appeared, she had a double gun-streak, false ports having been painted in her waist. Lieutenant Ballard, who had been carefully examining her with his glasses, remarked to the captain, who stood at his elbow, that she must at least be a fifty-gun ship. Stewart, after a long look, suggested that she was too small to be a ship of that class. "However," he continued, "be this as it may, you know I have promised you a fight before the setting of to-morrow's sun; and if we do not take it, now that it is offered, we may not have another chance. We must flog them when we catch them, whether she has one gun-deck or two."

Signals were now constantly interchanged between the vessels, and by three o'clock the "Constitution" had come so near that they were plainly made out to be two small frigates, or a frigate and a sloop-of-war, both close hauled on the starboard tack. The "Constitution," having the windward gauge, now manœuvred more carefully, and, hauling her sheets flat aft, pointed up so as to keep the advantage of position.

As the vessels came nearer and an action became certain, the stewards came on deck with the grog-buckets, in accordance with the

"NO 'DUTCH COURAGE' ON *THIS* SHIP"

time-honored rule on men-of-war by which the liquor is served before a fight. Instructions had been given that, as the battle was to be with two ships, a double portion of the drink should be served. But just as the stewards were about to ladle it out an old quartermaster rolled down from forward, and saying, "We don't want any 'Dutch courage' on *this* ship," with a great kick sent the bucket and its contents flying into the scuppers.

About four o'clock the westernmost ship signalled her consort and bore down to leeward to join her. The "Constitution" now set her stu'n-sails and went bearing down after them at a strain that seemed to menace her spars. She was rapidly drawing up with them when, just as she got well within range of the long guns, there was a sharp crack far aloft and the royal-mast snapped off at the cap. It was a doubtful moment, for the Englishmen crowded on all sail to escape, and rapidly drew together, flinging out their English ensigns as though in triumph.

But they did not reckon on the superb seamanship of the "Constitution." In a trice the men were aloft with their axes, the wreck was cleared away, new gear was rove, and in half an hour a new mast was aloft and another royal was spread to the breeze.

But the ships had been enabled to close with each other, and Stewart had lost the opportunity of attacking them separately. They made one ineffectual effort to get the weather-gage, but, finding that the "Constitution" outpointed them, they settled back in line of battle and cleared ship for action. Stewart immediately showed his colors and beat to quarters.

The fog had blown away and the sun had set behind a lowering bank of clouds. The wind still blew briskly, but the "Constitution" only pitched slightly, and offered a fairly steady platform for the guns, which were now trained upon the nearest vessel, but a few hundred yards broad off the port-bow. The darkness fell rapidly, and the moon came out from behind the fast-flying cloud-bank and silvered the winter twilight, gleaming fitfully on the restless water, a soft reproach upon the bloody work that was to follow.

At a few moments past six the long guns of the "Constitution's" port-battery opened fire, and the battle was on. Both ships responded quickly to the fire, and for fifteen minutes the firing was so rapid that there was not a second's pause between the reverberations. The English crews cheered loudly. But the gunners of the "Constitution" went on grimly with their work, sponging and loading as though at target-practice, content to hear the splintering of the timbers of the nearest vessel as the double-shotted thirty-twos went crashing into her. Before long the smoke became so thick that the gunners could not see their adversaries; and Stewart, ordering the batteries to cease firing, drew ahead and ranged abeam of the foremost ship, with his port-battery reloaded and double-shotted. He waited until he was well alongside before giving the order to fire, when he delivered such a terrible hail of round-shot, grape, and canister that the enemy staggered and halted like an animal mortally wounded. For the moment her battery was entirely silenced, and during the lull they could hear the cries of the wounded as they were carried below to the cockpit. The English cheered no longer. Another such a broadside might have finished her; but before Stewart could repeat it he saw that the other ship was luffing up so as to take a raking position under the stern of the "Constitution."

Nowhere did the wonderful presence of mind of Stewart and the splendid seamanship of his crew show to better advantage than in the manœuvre which followed. He quickly braced his main- and mizzen-topsails flat to the mast, let fly all forward, and actually backed down upon the other enemy, who, instead of being able to rake the "Constitution," found her emerging from the smoke abreast his bows in a position effectually to rake *him*. The "Constitution's" guns by this time had all been reloaded, and a terrific fire swept fore and aft along the decks of the Englishman, tearing and splintering her decks and dismounting many of the guns of both batteries. So terrible was the blow that she faltered and fell off. Before she could recover from the first, another terrible broadside was poured into her.

The other vessel now tried to luff up and rake the "Constitution" from the bows. But the American filled away immediately and let them have her other broadside. Side by side the "Constitution" and the larger ship sailed, firing individually and by battery as fast as they could sponge and load. Here and there a shot would strike within the stout bulwarks of the American; and one of these tore into the waist, killing two men and smashing through a boat in which two tigers were chained. A sailor named John Lancey, of Cape Ann, was carried below horribly mutilated. When the surgeon told him he only had a few moments to live, he said, "Yes, sir, I know it; but I only want to know that the ship has struck." Soon after, when he heard the cheers at her surrender, he rose from his cot, and, waving the stump of his blood-stained arm in the air, gasped out three feeble cheers and fell back lifeless.

Having silenced the larger vessel, Stewart immediately hurried to the smaller one, which had been firing through the smoke at the gun-flashes. The "Constitution" fell off, and, gathering headway, succeeded in getting again across her stern, where she poured in two raking broadsides, which practically cut her rigging to pieces. Returning to the larger vessel, Stewart rounded to on her port-quarter and delivered broadside after broadside with such a telling effect that at 6.50 she struck her colors.

The other vessel having in a measure refitted, came down gallantly but foolishly to the rescue of her consort. The "Constitution" met her with another broadside, which she tried to return, and then spread all sail to get away. But the American ship could outsail as well as outpoint her, and under the continuous fire of the bow-chasers of the "Constitution" she became practically helpless, and at about ten o'clock, when the dreaded broadside was about to be put into play again, she surrendered.

It was a wonderful battle. In a fight between one sailing-ship and two the odds were fourfold on the side of the majority. For it was deemed next to impossible to rake without being doubly raked in return. This obvious disadvantage was turned by Stewart to his own account by what critics throughout the world consider to be the finest manœuvring ever known in an American ship in action. He fought both his broadsides alternately, and luffed, wore, or backed his great vessel as though she had been a pleasure-boat. Neither of his adversaries succeeded in delivering one telling raking broadside. She seemed to be playing with them, and skilfully presented her reloaded guns to each vessel as it attempted to get her at a disadvantage.

The larger vessel was discovered to be the "Cyane," 34, Captain Gordon Falcon, and the smaller one the sloop-of-war "Levant," 21, Captain George Douglass. The "Constitution" had fifty-one guns, while the Englishmen had fifty-five; but of the "Constitution's" crew four were killed and ten wounded. On the "Cyane" and "Levant" thirty-five were killed and forty-two were wounded.

After the battle, while the two English captains were seated in Stewart's cabin dining with their victor, a discussion arose between them in regard to the part each had borne in the battle, while Stewart listened composedly. Their words became warmer and warmer, and each accused the other in plain terms of having been responsible for the loss of the vessels. At a point when it seemed as though the bitterness of their remarks bade fair to result in blows, Stewart arose and said, dryly,—

"Gentlemen, there is no use getting warm about it; it would have been all the same, whatever you might have done. If you doubt that, I will put you all on board again, and we can try it over."

The invitation was declined in silence.

For this gallant action Congress awarded Stewart a sword and a gold medal, and "Old Ironsides" soon after the war was over was temporarily put out of commission. Her day of fighting was over. But years after, refitted and remodelled, she served her country in peace as gracefully as she had served it gloriously in war.

THE "CONSTITUTION" AND THE "GUERRIERE"

BY the exercise of remarkable seamanship Captain Hull had succeeded in escaping from the British squadron, under Broke, off the Jersey coast. But he came so near capture that the secretary of the navy succeeded in frightening himself and the whole Cabinet at Washington into such a state of timidity that, had they had their way, no war-vessel flying the American flag would have been allowed to leave any Atlantic seaport and put to sea.

Captain Hull had carried the "Constitution" into Boston, where, if the orders had reached him in time, the secretary would have peremptorily bidden him to remain. But Hull was not in a humor to be inactive. What he wanted was a fight, yard-arm to yard-arm, with a frigate of the enemy, preferably the "Guerriere," Captain Richard Dacres, who had sailed boldly up and down the coast with an open challenge to any frigate flying the American flag. Though very warm personal friends ashore, both Hull and Dacres had high opinions of the merits of their own vessels. Dacres voiced the prevailing sentiment of the officers of his navy when he spoke of the "Constitution" as a bunch of pine boards which the British would knock to pieces in twenty minutes. Hull said little; but several months before war was declared had met Dacres, and wagered him a cocked hat on the result should the "Constitution" and the "Guerriere" ever meet. With the timidity at home, neither he nor any American officers had much encouragement. Although the Navy seemed popular with the public, until our vessels were triumphant the officers were not sure that their movements would always be approved by the Government.

But Hull knew he had a good ship and a good crew. He had trained them himself, and he knew what they could do aloft and at the guns. Moreover, he knew what he could do himself. The navy was small, but the men who had smelt powder in the Revolution and before Tripoli were a stalwart set and had done deeds of gallantry that had set the greatest admirals of Europe by the ears. Many ingenious contrivances had been adopted, to be now tried for the first time. Sights had been put upon the guns, and the gun-captains knew better how to shoot than ever before. So, without waiting for the orders from the secretary which he knew would hold him in port indefinitely, Hull sailed on the first fair wind and uncompromisingly put out to sea. If the orders came, he wouldn't be back to obey unless he had captured a British frigate, or, at the very least, some merchant prizes. If he *did not* succeed, it meant that he might be hung or shot for sailing without orders. But even this sword of Damocles did not deter him. He would do his best, at any rate, and made a quiet seaman's petition to the God of winds and seas to send him the "Guerriere."

Thinking to find a better opportunity towards Halifax, where many British men-of-war and merchantmen put in, Hull sailed to the northward, and cruised as far as the mouth of the Gulf of St. Lawrence. The frigate "Spartan," 38, was in those waters; but after watching for her for some days, he stood out to sea. On the 15th of August he sighted five vessels. The "Constitution" set all sail and rapidly came up with them. Four of them scattered, leaving the fifth, a brig, on fire. Hull made for the largest of the others, and found her to be an English merchantman in charge of an American prize-crew. The "Constitution" saved her from capture at the hands of the other vessels. Before night another vessel was overhauled, and she was found to be the American "Adeline," in the hands of a prize-crew from the British "Avenger." One vessel was destroyed and the other was sent to Boston in charge of Midshipman Madison and five men, carrying the first suggestion of the brilliant news which was to follow.

A few days later the "Constitution" chased and overhauled the American privateer "Decatur," which, believing her to be an English cruiser, had thrown overboard almost all of her guns. The captain of the privateer had good news, though. He had sighted an English frigate the day before, sailing southward under easy sail. Hull immediately set everything the "Constitution" could carry and gave the quartermasters a course which should enable him to come up with her by the following day.

The next morning dawned clear, but the breezes fell light, and not until the morning watch was there wind enough to send the American frigate bowling along on her course under top-gallant-sails and royals. Hull took the deck for awhile himself and sent lookouts to the fore- and main-royal-yards to keep a sharp lookout. With moderate luck they should catch up with her. And then Hull felt that he would make the "Constitution" the most talked about ship afloat or else he would change the timidity at the Navy Department into a panic for which there would be some reason.

If the ship were the "Guerriere," he promised himself a new hat.

Not a sail hove in sight until towards two in the afternoon, when a lookout aloft shouted, in a voice that was taken up by four hundred throats on the spar- and gun-decks,—

"Sail ho!"

In a moment the watch below came rushing up. So great was the excitement that many of them went half-way to the tops, without orders or permission, to view the stranger. In an hour the stronger glasses proved her plainly to be a frigate, and the "Constitution" eased off her sheets, and with the bit in her teeth boomed steadily down for her. For an hour the two ships moved in this position, the stranger making no effort to escape and leaving her colors, which were soon made out to be British, flying in defiance. In fact, as soon as she discovered the "Constitution" to be an American frigate she took in sail, laid her maintop-sail to the mast, and silently awaited the approach. Hull sailed on until within about three miles of the enemy, when he sent his light yards down, reefed his topsails, and cleared ship for action.

An American-built frigate was for the first time to test her stanchness against a worthy representative of the mistress of the seas and "Terror of the World." Most of the crew had never been in close action before. The chase of the "Constitution" had tired their hearts less than their bodies, for the firing of the British squadron had been at a very long range, and there was never a time when their ship was in danger from the cannonading of the enemy. There was not a qualm or a fear to be seen on the faces either of grizzled seaman or powder-boy, and they went to quarters with enthusiasm.

But underlying it all there was a note of gravity. They were going to bring an American ship into action with a frigate whose navy had scored hundreds of victories over the vessels of all the great nations of the earth. They half wondered at their audacity and that of their captain in defying a frigate so redoubtable as the "Guerriere," for there seemed no further doubt that it was she. But they looked up at Hull, who was calmly pacing up and down the quarter-deck, taking a look now and then at the enemy through his glass, and their confidence came back to them. The excitement was intense, and one by one the men began throwing aside their shirts and drawing in the buckles of their cutlass-hangers, most of the gun-crews stripping themselves to the waist and casting aside their shoes to avoid slipping on the decks when the blood began to flow. More than one of them had his own private score to settle with the British navy. Many of them had been at one time or another taken off American merchant-ships and impressed into the service of the enemy, and some of them still bore upon their backs the scars of the bloody lashes of the relentless "Cat." The father of Captain Hull had died in the pest-ship "Jersey," in the Revolution, and the other officers had all some grievances of their own which made them look eagerly forward to the battle which they intended should mean victory or death.

On the "Guerriere" there was a feeling of unshaken confidence. That any calamity to their ship could be expected from an American-built vessel, manned by a crew collected haphazard among the merchant-ships of the Atlantic harbors, never for a moment occurred to them. When the drum beat to quarters, the men tumbled to their stations willingly enough, with no more trepidation than if they were going to target-practice. Captain Dacres summoned an American prisoner, the captain of the captured merchant-brig "Betsy," and asked him what he thought of the vessel which was approaching. The skipper ventured that she was undoubtedly an American frigate. Captain Dacres replied with a smile,—

"She comes down a shade too boldly for an American." And then added, "Well, the better he behaves the more honor we will have in taking him."

As the "Constitution" bore down nearer, her ensign and jack flying proudly, there could no longer be any doubt as to her nationality and intentions, and he shouted to his crew, who stood at the guns,—

"There, my men, is a Yankee frigate. In forty-five minutes she is certainly ours. Take her in fifteen, and I promise you four months' pay."

Shortly after this Captain Hull was within two or three miles, and the "Guerriere" opened fire on the "Constitution," to try the distance and get the range.

The shots fell short, but Hull took in his light sails and came down more warily under topsails. The "Constitution" fired a broadside, but these shots, too, dropped in the water between them. As he came nearer, the "Guerriere" squared away, wearing first to port and then to starboard, firing alternate broadsides and manœuvring to avoid being raked. He wanted to cripple the American's rigging from a distance, if possible. But the shot all missed their mark, and the "Constitution" only replied with her bow-guns. Hull soon saw that this manœuvring might last the day out without coming to close quarters, so he hoisted his top-gallant-sails and made straight for the enemy.

Now the shot of the Englishman began coming aboard. Some of the standing rigging was cut away and the vessel was hulled several times. But the men, having carefully reloaded, stood silently at their guns, looking out through the ports at the "Guerriere," which, enveloped in smoke, kept up a continuous fire. They looked anxiously at the short, stout, sturdy figure of Captain Hull, but he continued pacing the quarter-deck, making no sign that he was aware of the damage the shots were causing. In a moment the report of "Nobody hurt yet, sir," ceased suddenly. A shot struck the "Constitution's" starboard bulwarks up forward and sent a jagged hail of splinters among the crew of two of the guns of the first division. Two men were killed outright and one or two more were wounded by this shot, and as their shipmates saw the men carried below to the cockpit they moved uneasily, and several of the gun-captains wished to fire. Lieutenant

Morris now, with a view to quieting them, strode aft to the quarter-deck, where Hull was still calmly pacing up and down, and said,—

"The enemy has killed two of our men. Shall we return it?"

"Not yet, sir," replied the imperturbable Hull.

Morris returned to his station. But there is nothing more disorganizing to men than to be fired at and not have the opportunity of firing in return, and they besought Morris again to give the permission. Twice more the lieutenant went aft to the quarter-deck, and twice he got the same reply. Hull, like Paul Jones, believed in great broadsides at close quarters. This silence under galling fire was the greatest test of discipline an American crew had ever had. For in the heat of battle a man forgets to be afraid. That the men stood to it, speaks well for Hull's training.

At last the "Constitution," which had been drawing closer and closer, drew up to a position about forty yards off the "Guerriere's" port-quarter, and Hull, waiting until his guns could all bear, stooped low, bursting his breeches from knee to waistband in the excitement of the moment, and gave vent to all the pent-up feelings of two hours in the hoarse order,—

"Now, boys, give it to them!"

It was a well-directed broadside.

The shots crashed along the line of bulwarks and sent showers of splinters flying over her spar-deck. The ships were so close together that the effect of those shots could be seen distinctly. Some of the splinters flew as high as the mizzen-top, and instantly the English cheering ceased and the shrieks and cries of the wounded rang out between the concussions. Dacres now, for the first time, must have realized how great the honor would be if he took the "Constitution."

Nor did the action promise any sign of being over in fifteen minutes. So well aimed were the American guns that in a short time the enemy's main-yard was shot away, and he was otherwise damaged severely both below and aloft. At a little after six a twenty-four pound shot went through the "Guerriere's" mizzen-mast, and, swaying a moment, over it fell to starboard, making a wreck and drag which impeded the Englishman's manœuvres. The seas pounded it against the sides of the ship and a hole was knocked under her stern, through which she began taking water badly. When the mizzen-mast fell, Hull threw off his hat, and shouted,—

"Hurrah, boys, we've made a brig of her!"

IN THE TOPS OF THE "CONSTITUTION"

One of the seamen shouted back,—

"We'll make a sloop of her soon, sir!"

And they did; for in a little while the foremast followed by the board. The wreck trailing in the water astern acted as a rudder to the "Guerriere," and she swung across the wind. The "Constitution" forged ahead, and crossing her bows, poured in a raking broadside. Then swinging round to port, she sent in another as effective as the first. The ships were very close together, and a fire from a burning gun-wad broke out in the cabin of the American ship. This was quickly put out, however, by Lieutenant Hoffman of the after-gun division.

Both captains now decided to board, and the men were massed on the decks as they could be spared from the guns for the purpose. Dacres was on the point of sending his men across his bowsprit, but, finding the jackies of the "Constitution" ready to receive him, changed his mind. The sharpshooters in the tops of both vessels were firing into the black masses of men, and every shot told. Lieutenant Morris, on the "Constitution," while attempting to take a few turns of rope around the bowsprit of the "Guerriere," received a bullet through the body. William S. Bush, the first lieutenant of marines, while standing on the taffrail ready to board, was shot through the skull by a British marine, and instantly killed. John C. Alwyn, the sailing-master, at the same time received a ball through the shoulder. Captain Hull climbed up on the rail, when a Yankee seaman, putting his arms around him, dragged him down and out of danger.

"Not with them swabs on," he said, pointing to Hull's big bullion epaulettes. He would have been a certain mark for one of the sharpshooters of the enemy.

At about this time the flag of the "Constitution," which had been nailed at the mizzen-truck, was shot down. But a young topman, named Hogan, shinned up the spar far aloft, and, though fired at repeatedly by the British marines, succeeded in replacing it amid the cheers of his companions.

On the "Guerriere" things were going badly. Captain Dacres had been shot in the back by one of the American marines, but he pluckily remained on deck. As the "Constitution" got clear again, the mainmast, rigging and spars of the "Guerriere," which had been repeatedly cut by American shot, went over with a crash, and she lay on the wave completely helpless. This was less than half an hour after the "Constitution" sent in her terrible broadside.

The American ship drew off to a short distance to repair her damages, and in less than an hour returned, and sent Lieutenant Read in a cutter to discover if Captain Dacres had surrendered.

Dacres's humiliation was complete, and he felt that further battle would only be the butchery of his own brave fellows.

Lieutenant Read hailed him to learn if he had surrendered.

"I don't know that it would be prudent to continue the engagement any longer."

"Do I understand you to say that you have struck?" asked Read.

"Not precisely; but I don't know that it would be worth while to fight any longer."

"If you cannot decide," said the American, "I will return aboard my ship and resume the engagement."

Dacres here called out hurriedly,—

"I am pretty much *hors de combat* already. I have hardly men enough to work a single gun and my ship is in a sinking condition."

"I wish to know, sir," demanded Read peremptorily, "whether I am to consider you as a prisoner of war or as an enemy. I have no time for further parley."

Dacres paused, and then said, brokenly, "I believe now there is no alternative. If I could fight longer I would with pleasure, but I—I must surrender."

When Dacres went up the side of the "Constitution" to surrender his sword he was treated in the manner befitting his rank by a generous enemy. Captain Hull assisted him to the deck, saying, anxiously,—

"Dacres, give me your hand; I know you are hurt." And when the Englishman extended his sword, hilt forward, in formal surrender, Hull said, magnanimously,—

"No, no; I will not have the sword of a man who knows so well how to use it. But"—and his eyes twinkled merrily—"but I'll thank you for that hat." He had not forgotten the wager, if Dacres had.

The transferring of prisoners was at once begun, for it was seen that the "Guerriere" was a hopeless hulk, not fit to take to port. When this was all completed and every article of value taken from her, she was blown up, and the "Constitution" sailed for Boston.

She arrived at an opportune time. For Detroit had been surrendered without firing a shot in its defence, and the American arms on the Canadian frontier had otherwise met with disastrous failure. The "Constitution," gaily dressed in flags, came up the harbor amid the booming of cannon and the wildest of excitement among the people. A banquet was given to the officers in Faneuil Hall, and from that time the American navy gained a prestige at home it has never since lost. Congress voted a gold medal to Captain Hull, silver ones to the officers, and fifty thousand dollars as a bonus to the crew.

The statistics of the fight are as follows:

The "Constitution" had fifty-five guns, the "Guerriere" forty-nine, sending shot weighing approximately seven hundred and six hundred pounds respectively. The "Constitution's" crew numbered four hundred and sixty-eight; that of the "Guerriere" two hundred and sixty-three. The "Constitution" lost seven killed and seven wounded, and the "Guerriere" fifteen killed and sixty-three wounded. All authorities acknowledge that, other things being equal, the discrepancy in metal and crews hardly explains the difference in the condition of the vessels at the end of the battle.

THE "WASP" AND THE "FROLIC"

THE American frigates "Constitution," "Constellation," and "United States" fought and won great battles where the metal and crews were equal or nearly equal, and proved beyond a doubt the advantage of American seamanship and gunnery over the British in the Naval War of 1812. But it remained for the little sloop-of-war "Wasp," Captain Jacob Jones, to add the final evidence of Yankee superiority. Her action with the "Frolic" was fought under conditions so trying that it fairly ranks with the great frigate actions of our naval history.

The "Wasp" was only about one-sixth the size of the "Constitution." She was about as big as the three-masted schooners which ply in and out of our Atlantic seaports to-day, and only carried one hundred and forty men. What she lacked in size she made up in personnel, and what she lacked in ordnance she made up in precision of fire. They must have been fine Jack tars and gallant fellows every one of them, for there was no chance for skulkers in that fight. The vessel could not have been handled or the guns served as they were with one man less.

It was off Albemarle Sound, in the rough end of a Hatteras gale, with a gun-platform which now rolled the gun-muzzles into the spume and then sent them skyward halfway to the zenith. It is a wonder that the gunners could hit anything at all; but almost every broadside told, and the hull of the "Frolic" was again and again riddled and raked fore and aft.

When the war broke out the "Wasp" was in European waters, carrying despatches for the government. She was immediately recalled, and in October, 1812, sailed from the Delaware to the southward and eastward to get in the track of the British merchantmen in the West India trade. On the 15th of October she ran into a gale of wind off the capes of the Chesapeake, and lost her jib-boom and two men who were working on it at the time. For two days and nights the little vessel tumbled about under storm-sails, but Captain Jacob Jones was one of the best seamen in the navy, and no further harm was done. On the night of the 17th the wind moderated somewhat, though the seas still ran high. At about half-past eleven a number of frigates were seen, and Captain Jones deeming it imprudent to bear down nearer until day should show him who the strangers were, sailed up to get the weather-gage and await the dawn. His forward rigging was disabled, and he had no wish to take chances with an enemy of greatly superior force.

The dawn came up clear and cold, and, as the darkness lifted, the crew of the "Wasp" could make out six fine merchantmen under convoy of a big brig. The brig was about the same size as the "Wasp," and it was seen that several of the merchantmen mounted from eleven to eighteen guns each. Nevertheless, Jones sent his topmen aloft, and in a trice he had his light yards on deck and his ship reefed down to fighting-canvas. The vessel was rolling her bows half under, but the guns were cast loose and the decks cleared for action. The brig, too, showed signs of animation. Her men went aloft at about the same time as those of the "Wasp," and soon she signalled her convoy to make all sail before the wind to escape.

The sea was so high that it was eleven o'clock before the vessels came within range of each other. Then on the English vessel the Spanish flag was run up to the gaff. But the

Americans nevertheless held on a course which would soon bring the ships together. There were enough Englishmen in those waters for Jones to take chances of her being one of the enemy. By half-past eleven the ships were within speaking-distance,—two or three hundred feet apart,—and Captain Jones mounted the mizzen-rigging, lifting his voice so that it might be heard above the shrieking of the wind and sea, and shouted through his trumpet,—

"What ship is that?"

For answer the Spanish flag came down with a run, the British ensign was hoisted, and a broadside was fired. Just then a squall keeled the Englishman over to leeward, and the "Wasp" having the weather-gage, the shots whistled harmlessly overhead and through the rigging. The Yankee ship responded immediately. The gunners had been trained in all weathers to fire as their own vessel was about to roll downward on the wave towards their adversary. By this means the shots were more sure to go low in the enemy's hull and to have the additional chance of the ricochet which would strike a glancing blow. They waited a second or so for this opportunity, and then sent their broadside of nine shots crashing through the hull of the "Frolic."

The tumbling of the vessel sent the guns rolling about, and the tacklemen needed all their strength and skill to hold the guns in for serving and out for firing. But they were in no hurry. They worked as slowly and as surely as possible, taking every advantage of the roll of the vessel, training and aiming deliberately, and then firing at will. The Englishmen sent in three broadsides to two of the Yankees. But they fired from the hollow on the upward roll of the vessel and most of their shots went high, scarcely one of them striking the hull of the "Wasp."

It is a wonderful thing to think even of these two little vessels, tossed about like billets of wood, the playthings of the elements, fighting a battle to the death with each other, ignoring the roaring of the sea and the hissing of the water which now and again seemed completely to engulf them in its foam. The waves came over the bows and waist of the "Wasp," flooding the decks, overturning buckets and making division-tubs a superfluity. Sometimes it dashed in at the leeward ports, dipping the handles of the sponges and rammers, and even burying the muzzles of the guns, which the next moment would be pointing at the main-truck of their adversary. The powder-boys, wet to the waist, stumbled over the decks with their powder-charges under their jackets, and, though buffeted about and knocked down repeatedly, kept the men at the guns plentifully supplied with ammunition.

Although the British were firing rapidly and the shots were flying high, they began doing great damage in the rigging of the American. A few minutes after the battle was begun a shot from the "Frolic" struck the maintop-mast of the "Wasp" just above the cap, and it fell forward across the fore-braces, rendering the head-yards unmanageable for the rest of the action. A few minutes later other shots struck the mizzen-top-gallant-mast and the gaff, and soon almost every brace was shot away. The "Frolic" had been hulled repeatedly, but aloft had only lost her gaff and head-braces. In a quiet sea it would have been bad enough to lose the use of the sails, but in a gale of wind manœuvring became practically impossible. The wind was blowing fiercely so both vessels drove on before it, keeping up the cannonading whenever a gun would bear, and pouring in from the tops a fire of musketry upon the officers and men upon the decks.

The "Wasp," having squared forward by the dropping of her maintop-mast across the fore-braces, no longer sailed on the wind, and in a moment drew forward, gradually approaching across the bows of the "Frolic," which, having lost the use of her head-sails, could not sheer off. Captain Jones was quick to see his advantage, and ran the enemy's bowsprit between the main- and mizzen-masts of the "Wasp." The vessels now began striking and grinding against each other furiously, as though by a test of the stanchness of their timbers to settle the battle between them. The men who were loading two of the port broadside guns of the "Wasp" struck the bow of the "Frolic" with their rammers and found themselves looking into the forward ports of the enemy. The guns were loaded with grape, and after the ships crashed together were fired directly through those forward ports of the "Frolic," raking her from stem to stern in a frightful manner.

The next wave tore the ships apart, and the "Wasp" forged ahead, the bowsprit of the Englishman catching in the mizzen-shrouds, where Lieutenant James Biddle and a party of officers and seamen were awaiting the order to board. In this position the bowsprit of the "Frolic" was pounding terribly upon the poop of the "Wasp." At every send of the waves the bows of the Englishman would fall as the stern of the American rose, and it seemed as though both ships would be torn to pieces. The men of the "Wasp" had wished to board, the moment the ships had come together, and crowded along the hammock-nettings hardly to be restrained. But Captain Jones, knowing the advantage of his raking position, wanted to send in another broadside. He called the men back to the guns, but they were too intent upon their object. One brawny fellow, named Jack Lang, who had been impressed into the British service, made a spring, and catching a piece of gear, swung himself up on the bowsprit and clambered down alone, his cutlass in his teeth, to the enemy's deck. The "Wasp's" men cheered vigorously, and, leaving their guns, rushed aft to follow him. Captain Jones, seeing that they would not be denied, then gave the order to Lieutenant Biddle to board.

Biddle, cutlass in hand, jumped upon the nettings to lead the men. Midshipman Yorick Baker, being too small to clamber up alone, and seeing Biddle's coat-tails flapping in the wind, seized hold of them, one in each hand. He did not want to be left behind, and thought he might trust to the impetuosity of his superior officer to land him successfully among the first on the deck of the enemy. But just then a terrific lurch threw Biddle off his balance, and they both came violently to the deck. They were up again in a second, however, and with Lieutenant George W. Rogers and a party of seamen finally reached the bowsprit of the "Frolic."

Upon the fo'c's'le of the enemy stood Jack Lang, swinging to the motion of the brig, his cutlass at his side, looking aft at a scene of carnage that was hardly imaginable. All the fierceness had died out of him, for he looked around at Biddle and grinned broadly. The decks were covered with the dead and dying, who tossed about in the wash of bloody water with every heave of the ship. The decks, masts, bulwarks, and rails were torn to ribbons, huge jagged splinters projecting everywhere. Guns, tubs, sponges, rammers, and solid shot were adrift, pounding from one side of the wreck to the other. No one moved to secure them, for only half a dozen men stood upright. At the wheel an old quartermaster, badly wounded, swung grimly, ready to die at his post. Behind him an English lieutenant, bleeding from ghastly wounds, clutched at a stanchion for support. Two other officers stood near, and one or two jackies glared forward at the Americans. There was no sign of resistance, and the wave of pity which came over Biddle and his officers swept away all desire for battle. The British flag was still flying. No one seemed to have the strength to haul it down; so Biddle went aft and lowered it to the deck. In a few moments the masts fell, and she lay a useless hulk wallowing upon the waves, which, more sure of their prey, dashed against her torn sides, widening the gashes made by her indomitable enemy, and at times making clean breaches over her bulwarks, tearing loose her boats and otherwise completing her destruction.

Under the conditions, it seemed hardly credible that such injury could have been inflicted in so short a time, for the battle had lasted only forty-three minutes. The "Frolic" had twenty-two guns, while the "Wasp" had only eighteen. The crew of the "Frolic" was less than of the "Wasp," the best authorities estimating it at one hundred and ten, against one hundred and thirty-eight of the "Wasp." But even here the great loss and damage to the "Frolic" can be explained in no way save that the Americans were superior gunners and seamen. The "Wasp" lost five killed and five wounded, and these men were most of them shot while aloft trying to refit gear. The "Frolic" lost fifteen killed and forty-seven wounded, making a total of sixty-two against ten of the "Wasp."

But Jacob Jones's victory was not to prove profitable, save in the great moral influence it exercised in England and America. He placed a crew upon the prize, and, having cleared away his wreck and refitted his rigging, tried to make sail away after the fleet of merchantmen, which by this time were nearly hull down on the horizon. But a great British seventy-four, the "Poictiers," hove in sight, and before Jones could get away he found himself under her guns a prisoner. Captain Beresford, of the line-of-battle ship, took the sloop-of-war to Bermuda, and there a garbled report of the action between Captain Whinyate's and Captain Jones's vessels was written. But the American captain and his gallant crew were soon exchanged, and returned home, where their victory had been given its true value. They received twenty-five thousand dollars from Congress as prize-money, and a gold medal was given to Captain Jones and a silver one to each of the officers. The legislature of Pennsylvania gave Lieutenant Biddle a sword for his gallantry.

THE "CONSTITUTION" AND THE "JAVA"

AT the beginning of the war of 1812 there were but three first-class frigates in our navy, and but five vessels of any description were fit to go to sea. But the war with Tripoli and the gallant deeds of the American officers had made the service popular with the public. In March, 1812, an act was passed which appropriated money to put all these vessels in condition to meet the enemy on a more equal footing, and a naval committee was formed to deal with the emergency. Langdon Cheves was appointed chairman, and he took hold of the great task of rebuilding and regenerating the naval service with enthusiasm and good judgment. The result was that the committee expressed the opinion "that it was the true policy of the United States to build up a navy establishment, as the cheapest, the safest, and the best protection to their sea-coast and to their commerce, and that such an establishment was inseparably connected with the future prosperity, safety, and glory of the country."

When wār was declared, the "Constitution" was in good condition, but the "Chesapeake" and the "Constellation" were not seaworthy. These were recommended to be immediately put in condition, and ten other frigates, averaging thirty-eight guns each, to be built. There was no difficulty in raising the crews for these vessels. Owing to the impressment of American and other seamen into the British service, the Cross of St. George had come to be so hated by the fishermen, coastwise sailors, and merchantmen that they sailed, drove, or walked to Philadelphia, Baltimore, and the other places where the frigates were fitting out, eager to sign the articles which made them American men-o'war's-men. They were not drafted into the service like many of the British Jackies, at the point of the pistol, but came because they wanted to, and because with the building up of a new navy there came a chance to see the flag they hated trailed in defeat. That and nothing else was the reason for the wonderful success of American arms upon the sea during the war of 1812. The American officers, smarting under past indignities to the service and to themselves, went into the many actions with determination and enthusiasm, combined with the experience of a rough-and-tumble sea, —experience which with anything like an equal force meant either victory or absolute destruction.

The "Constitution," under Hull, had escaped from the British squadron, under Broke, off the Jersey coast, had defeated the frigate "Guerriere," and in all her history had shown herself to be a lucky ship. William Bainbridge had been given the command of the "Constellation," but, arriving at Boston, Hull had found it necessary to give up his command, and Bainbridge immediately applied for "Old Ironsides."

The victories of the American frigates "Constitution" and "United States" over the British "Guerriere" and "Macedonian" had aroused great enthusiasm throughout the country, and the government had decided to change its timorous policy. Hoping to draw some of the British vessels away from the coast and cause them to be distributed over a wider horizon, expeditions were arranged to strike the enemy at many distant points. Bainbridge's orders were to sail for the Indian Ocean and capture or destroy as many English merchant-vessels as possible. His squadron, besides the "Constitution," 44, consisted of the "Essex," 32, Captain David Porter, and the "Hornet," 18, Master-Commandant James Lawrence. Bainbridge and Lawrence put to sea from Boston on the 26th of October, while Porter left the Delaware on the 24th to rendezvous at Porto Praya, on the South American coast.

A few days later, H. M. S. "Java," a thirty-eight-gun frigate, Captain Henry Lambert, having two merchant-ships under convoy, sailed from Portsmouth, England, for India. She also had as passengers the newly-appointed governor of India, Lieutenant-General Thomas Hislop, and many naval and army officers, who were being carried out to their posts.

The "Constitution," arriving at Porto Praya, and failing to find Porter in the "Essex," put to sea again, stopping at Fernando de Noronha in the hope of meeting her there. Lawrence, in the "Hornet," challenged the British sloop-of-war "Bonne Citoyenne" to single combat; but her commander declined, in view of the presence of the "Constitution." Bainbridge wrote that he would not interfere, and pledged him his honor to give the Englishman the opportunity to fight the "Hornet" to the death. Hoping to bring the action about, Bainbridge sailed away, and remained four days. But the British captain was determined not to fight, and Lawrence was thus denied the opportunity he afterwards had with the ill-fated "Chesapeake."

"CONSTITUTION" AND "JAVA"

Near the end of December, 1812, the "Constitution" was cruising off the coast of Brazil, about thirty miles from Bahia. The wind was light from the northeast, and Bainbridge was moving under short sail. "Old Ironsides," a ready sailer when in condition, had been off the stocks so long and was so befouled by her stay in tropical waters that she moved rather sluggishly, and had not the capacity for legging it that she had when Hull had carried her from under the guns of the British squadron. Her sails were patched and her rigging was old, but Bainbridge had done all he could with her, and his men were full of confidence. She was the "Constitution," and that was enough for them. They only wanted an opportunity to repeat or surpass some of her previous exploits.

They had not long to wait. At nine o'clock on the morning of December 29, the man at the fore-crosstrees passed the cry of "Sail-ho," and soon from the deck two sails could be seen to the north, near the coast. They were both made out to be full-rigged ships, one standing in cautiously for the land and the other keeping a course out to sea, pushing down gallantly under a full press of canvas. The one inshore was the American ship "William," which had been captured by the British, and the other was the "Java." The jackies who lined the nettings of the "Constitution" soon discovered that their wishes were to be granted, for the larger ship was evidently determined to come up, and could be nothing but a man-of-war looking for a fight.

By about eleven Captain Bainbridge took in his royals and went about on the other tack. The Englishman was coming nearer now, and hoisted the private signals, English, Spanish, and Portuguese, in succession. Bainbridge hoisted the private signal of the day, and finding that it was not answered, cleared ship for action immediately. Then, wishing to draw his enemy from his consort, he set his mainsails and royals and stood out to sea. The "Java" came up rapidly, and made sail in a parallel course. Finding that the other ship did not follow, and desiring to make the other vessel disclose her identity, Bainbridge showed his colors,—his broad pennant at the main, the Stars and Stripes at the peak, another at the maintop-gallant-mast, and the American jack at the fore. This was shortly followed by his adversary, who hoisted an English ensign and displayed a private signal.

All this time the "Java" was rapidly gaining on the "Constitution," and Bainbridge, finding that he was outsailed, took in his royals and went about on the other tack, so as to pass within pistol-shot of the other.

The "Constitution," still a mile to leeward, soon fired a shot across the "Java's" bows to induce her to show her colors, which she had hauled down again. This had the desired effect, for the bits of bunting went up with a run, and a whole broadside was fired at the "Constitution." But the range was too great for successful marksmanship, both these shots and those fired by the "Constitution" in return dropping harmlessly alongside.

By a little after two o'clock the frigates were within half a mile of each other, and the action then began with great spirit. The Englishman got the range first, and sent in a broadside which hulled the "Constitution" and killed and wounded several of her men. It soon became evident to Bainbridge that Captain Lambert's guns carried better than his own, so began luffing up repeatedly in order to shorten the distance for an effective broadside. He was sure of his marksmanship if once his men got the range, for the same gun-captains were with him that had helped Hull to her great victory over the "Guerriere." It was difficult to draw up, as the Englishman was forging ahead with the evident desire to sail close to the wind and keep the weather-gage at all hazards. The "Constitution" could only luff up at opportune moments, for Lambert's position was one which would enable him to rake the "Constitution" from stem to stern if he luffed when the broadside was ready. But he edged up cautiously, and soon the vessels were but musket-shot apart. A continuous fire now began, and the wind being light, both vessels were soon so shrouded in smoke that only at intervals could the gunners make out their adversaries. Along they sailed, side by side, giving and receiving tremendous volleys. About this time a solid shot went crashing along the quarter-deck of the "Constitution" and, striking her wheel, smashed it to pieces. The gear had been rove below, however, and the ship throughout the remainder of the battle was steered by means of tackles on the berth-deck. The captain's orders were shouted down through the after-hatch and repeated by a line of midshipmen to the men at the tackles.

Bainbridge, in full uniform, stood by the weather-rigging at the time the disabling shot came aboard, and a small copper bolt drove through the upper part of his leg, inflicting a bad wound. But fearing that if he left the deck his men might lose some of the ardor with which they were fighting, he would not go below though frequently urged so to do. Instead of this he bound it up with his handkerchief, and remained at his post, his epaulettes a fair mark for the sharpshooters in the tops of the enemy. His men down in the waist of the "Constitution" looked now and again at the imposing figure by the mizzen-mast, and bent to their work with a will, firing as rapidly as their guns could be loaded. The distance between the ships was now so short that all the smaller guns and carronades could be used, and a rapid and well-directed fire was kept up both upon the hull and the spars of their adversary.

The "Java," by her superior sailing qualities, was enabled to reach well forward on the "Constitution's" bow when she eased off her sheets to round down across the bows of the American and rake. But Bainbridge, in spite of the disadvantage of wrecked steering-gear, was too quick for her. He put his helm up, and wore around in the smoke, thus keeping his broadside presented. The Englishman at last succeeded in getting under the "Constitution's" stern and pouring in a broadside at close range. But, fortunately, comparatively little damage was done. The superiority of the gunnery of the Americans, save for a few of the Englishman's well-directed shots, had been from the first far superior to that of the

Englishmen. The fire of the "Java" was far less rapid and less careful than that of the "Constitution." Had the gunnery been equal, the story of the fight would have had a different ending.

But the Americans labored under a great disadvantage, and Captain Bainbridge, determined to close with the enemy at all hazards, put his helm down and headed directly for the enemy, thus exposing himself to a fore-and-aft fire, which might have been deadly. But for some reason the Englishman failed to avail himself of this opportunity, only one 9-pounder being discharged. When near enough, the "Constitution" rounded to alongside and delivered her entire starboard broadside, which crashed through the timbers of the "Java" and sent the splinters flying along the entire length of her bulwarks. The shrieks of the injured could be plainly heard in the lulls in the firing, and soon the bowsprit and jib-boom of the enemy were hanging down forward, where they lay, with the gear of the head-sails and booms in a terrible tangle. With this misfortune the "Java" lost her superiority in sailing, and this was the turn in the action. Quickly availing himself of this advantage, Bainbridge again wore in the smoke before Captain Lambert could discover his intentions, and, getting under the "Java's" stern, poured in a rapid broadside, which swept the decks from one end to the other, killing and wounding a score of men. Then sailing around, he reloaded, and fired another broadside from a diagonal position, which carried away the "Java's" foremast and otherwise wrecked her.

Captain Lambert, now finding his situation becoming desperate, determined to close with the "Constitution" and board her. He tried to bear down on her, but the loss of his head-yards and the wreck on his forecastle made his vessel unwieldy, and only the stump of his bowsprit fouled the mizzen-chains of the American vessel. The American topmen and marines during this time were pouring a terrific fire of musketry into the mass of men who had gathered forward on the English vessel. An American marine, noting the epaulettes of Captain Lambert, took deliberate aim, and shot him through the breast. Lambert fell to the deck, and Lieutenant Chads assumed the command. The Englishmen, disheartened by the loss of their captain, still fought pluckily, though the wreck of the gear forward and the loss of their maintop-mast seriously impeded the handling of the guns. At each discharge their sails and gear caught fire, and at one time the "Java's" engaged broadside seemed a sheet of flame. At about four o'clock her mizzen-mast, the last remaining spar aloft, came down, and she swung on the waves entirely dismasted. It seemed impossible to continue the action, as but half a dozen guns could be brought to bear.

The "Constitution," finding the enemy almost silenced and practically at her mercy, drew off to repair damages and re-reeve her gear. Bainbridge had great confidence in the *look* of the "Constitution," as, to all outward appearances unharmed, she bore down again and placed herself in a position to send in another broadside. His surmise was correct, for the one flag which had remained aloft was hauled down before the firing could be resumed.

Lieutenant George Porter, of the "Constitution," was immediately sent aboard the Englishman. As he reached the deck he found the conditions there even worse than had been imagined by those aboard the "Constitution." Many of the broadside guns were overturned, and, though the wreck had been partially cleared away, the tangle of rigging was still such that the remaining guns were practically useless. The dead and wounded literally covered the decks, and as the lieutenant went aboard the dead were being dropped overboard. The loss of her masts made her roll heavily, and occasionally her broadside guns went under. Lambert was mortally wounded. Lieutenant Chads, too, was badly hurt. When he had assumed command, in spite of the fact that he knew his battle was hopeless, he had tried to refit to meet the American when she came down for the second time. He only struck his colors when he knew that further resistance meant murder for his own brave men. The "Java" was a mere hulk, and the hulk was a sieve.

Comparison of the injuries of the "Java" and "Constitution" is interesting. With the exception of her maintop-sail-yard, the "Constitution" came out of the fight with every yard crossed and every spar in position. The injuries to her hull were trifling. The "Java" had every stick, one after another, shot out of her until nothing was left but a few stumps. It might have been possible to have taken her into Bahia, but Bainbridge thought himself too far away from home; and so, after the prisoners and wounded had been removed to the "Constitution," a fuse was laid, and the American got under weigh. Not long after a great volume of smoke went up into the air, and a terrific explosion was heard as the last of the "Java" sunk beneath the Southern Ocean.

When the "Constitution" arrived at Bahia, Captain Lambert was carried up on the quarter-deck, and lay near where Bainbridge, still suffering acutely from his wounds, had been brought. Bainbridge was supported by two of his officers as he came over to Lambert's cot, for he was very weak from loss of blood. He carried in his hand the sword which the dying Englishman had been obliged to surrender to him. Bainbridge put it down beside him on his bed, saying,—

"The sword of so brave a man should never be taken from him."

The two noble enemies grasped hands, and tears shone in the eyes of both. A few days afterwards the Englishman was put on shore, where more comfortable quarters were provided for him, but he failed rapidly, and died five days after.

The news of the capture of the "Java" created consternation in England. The loss of the "Guerriere" and the "Macedonian" were thought to have been ill-luck. But they now discovered an inkling of what they rightly learned before the war was over,—that the navy of the United States, small as it appeared, was a force which, man for man and gun for gun, could whip anything afloat.

When Bainbridge arrived in Boston he and his officers were met by a large delegation of citizens, and many festivities and dinners were held and given in their honor. The old "Constitution," rightly deserving the attention of the government, was put in dry-dock to be thoroughly overhauled. Of the five hundred merchantmen captured by Americans, she had taken more than her share, and of the three frigates captured she had taken two.

THE LAST OF THE "ESSEX"

WHEN Captain David Porter in the "Essex" failed to meet Captain Bainbridge in the "Constitution" off the Brazilian coast, and learned that the latter had captured the "Java" and returned to the United States, he was free to make his own plans and choose his own cruising-ground.

He captured an English vessel or so, but his ambition was to make a voyage which would result in the capture of as many vessels as could be manned from the "Essex." He thought the matter over at length and then formulated a plan which few other men would have thought of. No large war-vessel of the American government had been in the South Pacific for some years, and now the English whalers and merchantmen pursued their trade unmolested, save by a few privateers which sailed haphazard in the waters along the coast. David Porter decided to round the Horn, thus cutting himself off from his nearest base of supplies, and live the best way he might off vessels captured from the enemy.

He knew that he could not hope for a hospitable reception at any port he visited, but if he could keep his magazine and store-rooms supplied, determined to capture or destroy every vessel flying the British flag in those waters.

He started on his long voyage at the end of January, 1813, during the Southern summer season, when the gales and hurricanes in that region are at their fiercest. He had not been at sea very long before the scurvy broke out on the ship, and it was only by the most rigorous discipline and cleanliness that the disease was kept under control. By the middle of February the "Essex" reached the Cape, and, the weather having been moderately free from squalls, they were congratulating themselves on avoiding the usual dangers of those waters when a storm came up which in a short time began to blow with hurricane force. Gale succeeded gale, followed by intervals of calm, but nothing terrifying occurred until towards the end of February, when a storm which exceeded all the others in its fierceness began to blow. They were near a barren country, and, even should they reach land, there was no possible chance of escaping the slow torture of death from hunger and thirst. Great gray waves, measuring hundreds of feet from crest to crest, swept them resistlessly on towards the menacing shore, which could be seen dimly through the driving spray frowning to leeward. Many of the waves broke clear over the little frigate, knocking in her ports, opening her timbers, battering her boats to pieces as they swung on the davits, and loosening her bowsprit and other spars so that they threatened at each movement to go by the board. The crew, weakened and disheartened by disease and the excess of labor, lost heart and considered the "Essex" a doomed ship. David Glascoe Farragut, then a midshipman aboard of her, afterwards wrote that never before had he seen good seamen so paralyzed by fear at the mere terrors of the sea. On the third day an enormous wave struck her fairly on the weather-bow and broadside, and she went over on her beam ends, burying her lee-bulwark in the foam. It looked for a moment as if she would never right herself. The ports on the gun-deck were all stove in and she seemed to be filling with water. The head-rails were swept away, and one of the cutters was lifted bodily from the davits and smashed against the wheel. The fellows there stood bravely at their posts, though thoroughly terrified at the position of the ship. The water poured down below, and the men on the gun-deck thought she was already plunging to the bottom. The grizzly boatswain, crazy with fear, cried out in his terror,—

"The ship's broadside is stove in! We are sinking!"

That was the greatest of their dangers, though, and better days were in store for them. Early in March the "Essex" succeeded in reaching Mocha Island, and the men, starved on half and quarter rations, were sent ashore to hunt wild hogs and horses. These were shot in numbers and salted down for food. The crew soon regained their health and spirits, and Porter sailed away for Valparaiso, putting in there to refit his damaged rigging and spars.

And now began a cruise which is numbered among the most successful in the country's history. Porter had been at sea but a few days when he overhauled a Peruvian privateer, the "Nereyda." To his surprise, twenty-four American sailors were found prisoners aboard of her. When asked to explain, the Peruvian captain replied that as his country was an ally of Great Britain, and that as war was soon to be declared between Spain and America, he thought he would anticipate matters and be sure of his prizes. Porter, in forcible English, explained the Peruvian's mistake, and, to make the matter more clear, threw all his guns and ammunition overboard, so that he might repent of his folly in a more diplomatic condition.

The Peruvian captain begrudgingly gave Porter a list of all the English vessels in those waters. The first one captured was the whaler "Barclay." On the 29th of April the "Essex" took the "Montezuma," with a cargo of fourteen hundred barrels of whale-oil. Later in the same day the "Georgiana" and the "Policy" were overhauled. These prizes, with their cargoes, in England were worth half a million dollars; but, better than money, they were plentifully supplied with ropes, spars, cordage, stores, and ammunition, of which Porter still stood badly in need.

Finding that the "Georgiana" was a fast sailer and pierced for eighteen guns, Porter decided to make use of her as a cruiser, and, fitting her up, placed Lieutenant Downes in command of her, with forty men for a crew. Then the "Essex" took the "Atlantic" and the "Greenwich." With this very respectable squadron Porter sailed for the mainland, Lieutenant Downes in the "Georgiana" meanwhile capturing without great difficulty the "Catharine" and the "Rose." A third vessel, the "Hector," fought viciously, but was eventually secured after a stiff little battle.

Young Farragut had been made the prize-master of the "Barclay." He was only twelve years old, but Captain Porter, who was very fond of him, was confident of his ability to bring the ship into port. The English captain had been persuaded to act as navigator; but once out of sight of the squadron he refused to sail for Valparaiso. He afterwards said it was merely to frighten the boy. But the boy did not frighten at all. Instead he called one of his best seamen to him and ordered sail made. Then he told the captain that if he did not go below and stay there he would have him thrown overboard. The Englishman retreated below precipitately, and Farragut brought the ship safely in, a first proof of the courage and skill he was to show in after-life. Few boys of twelve would have done it even in those days when midshipmen soon became men regardless of age.

The "Atlantic," being reckoned the fastest vessel of her kind afloat in those waters, was now given to Downes, who had been promoted to master-commandant, and renamed the "Essex Junior." She was given twenty guns and sixty men, and soon proved her worth. All of this time Porter had been self-supporting. Neither he nor his squadron had cost his government a penny in money, and the prizes he captured, including the "Charlton," "Seringapatam," "New Zealand," and "Sir Andrew Hammond," could not be reckoned much short of a million and a half of dollars, a tremendous sum in those days, when the pay of a captain of a naval vessel was only twelve hundred dollars,—less than the pay of a boatswain to-day.

But Porter grew tired of his easy victories over merchantmen and privateers. He had succeeded in frightening the ships of the British entirely from the ocean. His one ship, a small frigate, had complete control in the South Pacific, and the Admiralty wondered at the skill and ingenuity of a man who could manage his fleets so adroitly. They determined to capture him; and two smart ships, the "Phœbe" and the "Cherub," were sent out for this purpose. Porter heard of their coming, and was willing enough to meet them if it were possible. He went to Nukahiva, in the Marquesas Islands, to put the "Essex" in thorough repair and give his men a rest. He remained there two months, sailing near the end of the year 1813 for Valparaiso, with the hope of their meeting the English cruisers.

The "Essex" had been there but a month when the "Essex Junior," which was cruising in the offing in anticipation of the arrival of the British ships, signalled, "Two enemy's ships in sight." Half the crew of the "Essex" were ashore enjoying sailor-men's liberty. Even if they all got aboard, it was fair to assume that they would be in no condition to fight should the Englishmen choose to violate the neutrality of the port by firing on them. Porter immediately fired a gun and hoisted the recall signal for all boats and men to return. The English captain, Hillyar, ran the "Phœbe" on the wind straight for the "Essex," the "Cherub" following closely. But when they reached the anchorage, the "Essex" was ready for action and the crew were at their stations. The "Phœbe" went around under the quarter of the "Essex," luffing up scarcely fifteen feet away. It was an exciting moment. Hillyar could see the men at their guns, and his ardor was perceptibly diminished. Had he given the order to fire then, he would have been raked fore and aft, and the tale of this last fight of the "Essex" might have had a different ending.

As it was, he jumped upon the nettings, and said, with distinguished politeness,—

"Captain Hillyar's compliments to Captain Porter, and hopes he is well."

Porter *was* well, but he was in no humor to bandy compliments.

"Very well, I thank you," he replied; "but I hope you will not come too near, for fear some accident might take place which would be disagreeable to you." And at a wave of his hand the kedge-anchors and grappling-irons were swung up to the yard-arms, ready to be dropped on the decks of the enemy. The men swarmed along the nettings, ready to jump aboard the Englishman as soon as she was close enough.

But Hillyar, not liking the looks of things, changed his tone considerably. He backed his yards hurriedly, and said in an excited manner,—

"I had no intention of getting aboard of you. I assure you that if I fall aboard it will be entirely accidental."

"Well," said Porter, "you have no business where you are. If you touch a rope-yarn of this ship I shall board instantly."

Porter then hailed Downes on the "Essex Junior" and told him to be prepared to repel the enemy. The vessels were in a position to be almost at the mercy of the Americans. When the "Phœbe" ranged alongside, the crews could see each other through the ports, and laughed and made grimaces at one another. One young fellow in the "Essex," who had come aboard drunk, stood at one of the guns, match in hand. He saw one of the English jackies grinning at him. He was primed for a fight, and yelled across,—

"I'll stop your making faces, my fine fellow." He leaned forward to apply the match to the vent, and was only saved from firing it in time by Lieutenant McKnight of the gun-division, who knocked him sprawling. Had that gun been fired, the "Phœbe" would have been taken.

There seems no doubt of Captain Hillyar's previous intention to try to take the "Essex" as she lay, regardless of the neutrality. Captain Porter would have been justified if he had fired at that time.

But the Englishmen were willing to bide their time. Two more British ships were expected, and they felt sure of their prey.

A strange state of affairs now ensued. The officers meeting on shore exchanged the proper courtesies, and strict orders were issued to the men, who for a wonder were restrained from fighting. Porter flew from his foremast a great white burgee, bearing the legend, "Free Trade and Sailors' Rights." Captain Hillyar soon hoisted one in reply, "God and Country: British Sailors' Best Rights. Traitors Offend Both." Porter then had another painted, and sent it to the mizzen, which read, "God,

Our Country, and Liberty. Tyrants Offend Them."

These amenities had the effect of making the crew eager for a speedy settlement of the question. Once Captain Hillyar fired a gun in challenge; but upon Porter's accepting it, the Englishman sailed down to his consort the "Cherub," and Porter returned. The Englishman, in spite of his challenge, was not willing to fight a single battle.

Finally, Captain Porter, learning of the expected early arrival of the "Tagus," 38, the "Raccoon," and two other ships, determined to put to sea and there fight it out with the two frigates as best he might. The next day, the 28th of March, 1814, a squall came up, and the "Essex" lost one of her anchors and dragged the other out to sea. Not a moment was to be lost in getting sail on the ship, for he saw a chance to sail between the southwest point of the harbor and the enemy. Under close-reefed topsails Porter made a course which seemed likely to carry him just where he wanted to go, when a heavy squall struck the ship, carrying away the maintop-mast and throwing the men who were aloft on the top-gallant-yard into the sea.

This great misfortune at a time when there

was at least a fighting chance of getting away put a different aspect upon the chances of the "Essex." Both English vessels immediately gave chase, and Porter, failing to make his anchorage, ran for shore, to anchor there and fight it out to the last drop of blood. The "Phœbe" and the "Cherub," bedecked with flags, came booming down to where Porter awaited them, flying flags from the stumps of his maintop-mast and at almost every point where he could run a halyard.

At about four o'clock the "Phœbe" selected a position under the stern of the "Essex," and opened fire at long range. The "Cherub" stood off her bow. The fire of the "Phœbe" was terribly destructive, and few guns from the "Essex" could be brought to bear upon her. The "Cherub" fared differently; and, finding her position too hot, sailed around and took up a position by her consort, where a tremendous fire was poured in. Captain Porter, with great difficulty, had three of his long 12-pounders hauled into his after-cabin, and at last succeeded in opening such a fierce and well-aimed fire that the enemy wore about and increased the distance between them. The "Phœbe" had three holes in her water-line, had lost the use of her mainsail and jib, and had her fore-main- and mizzen-stays shot away. Her bowsprit was badly wounded, and she had other injuries below.

But the "Essex" was fighting against terrible odds. The springs on her cables were again and again shot away and the crew were being killed and wounded in great numbers. When the ships of the enemy returned and opened a galling fire from such a position that it could not be returned by the "Essex," Porter determined to assume the aggressive. But when he attempted to make sail on his ship, he found that most of the running-gear had been cut away, only his flying-jib could be spread to the winds. But, nothing daunted, he cut his cable, and, spreading his tattered canvases the best way he could, made down for the "Cherub" until within range of the cannonades, where he gave the Englishman such a drubbing that he took to his heels and got out of range altogether. The "Phœbe" managed to keep her distance, and with her long guns kept sending in broadside after broadside, which swept the decks of the doomed "Essex" and mowed her men down like chaff. Captain Hillyar was taking no chances.

The slaughter on the "Essex" was horrible. One gun was manned by three crews, fifteen men being killed at it. Men were dying like sheep; but those who remained at the guns, and even the wounded, had no thought of surrender. A sailor named Bissley, a young Scotchman by birth, lost his leg. He lifted himself, and said to some of his shipmates,—

"I hope I have proved myself worthy of the country of my adoption. I am no longer of any use to you or her; so good-by." And before he could be restrained he pushed himself through the port into the sea and was drowned.

Midshipman Farragut acted as captain's aid, quarter-gunner, powder-boy, and anything that was required of him. He went below for some primers, when the captain of a gun was struck full in the face by a sixteen-pound shot, falling back upon the midshipman, spattering him with blood and tumbling then both down the hatch together. The blow stunned the midshipman for a moment; but when he recovered, he rushed again on deck. Captain Porter, seeing him covered with blood, asked him if he were wounded.

"I believe not, sir."

"Then, where are the primers?"

This first brought him completely to his senses. He rushed below again and brought the primers up. Captain Porter fell, stunned by the windage of a shot, but got to his feet unaided.

Though most other men would have surrendered the ship, Porter made up his mind to run her towards the shore and beach her broadside on, fight until the last and then blow her to pieces. An explosion occurred below and a fire broke out in two places. The decks were so covered with dead and dying that the men who remained upright could scarcely move among them. The cockpit would hold not another wounded man, and the shots which came in killed men who were under the surgeon's knife. Out of the two hundred and fifty-five souls who began the fight only seventy-five, including officers and boys, remained on the ship fit for duty. Many of the men, thinking the ship was about to blow up, had jumped overboard and had drowned or were struggling in the water in the attempt to swim to land. The long-range shots of the enemy were striking her at every fire. The Englishmen had the distance accurately and were battering her to pieces as though at target-practice.

Captain Porter, at last seeing that resistance was only a waste of life, called his officers into consultation. But one, Lieutenant McKnight, could respond, and at 6.20 P.M. the order was given to haul down the flag.

When the British boarding-officer came over the side, the sight of the carnage was so shocking that he had to lean against a gun for support. The force of the "Essex" was forty-six guns and two hundred and fifty-five men. That of the English, in conservative estimates, was seventy-three guns and four hundred and twenty-one men. The English lost five killed and ten wounded. The "Essex" fifty-eight killed, sixty-six wounded, and thirty-one missing.

Thus died the "Essex" in one of the bloodiest and most obstinate combats on record.

THE CAPTAIN OF THE MAINTOP

JAMES JARVIS was one of the "young gentlemen" on the "Constellation" during the war with France. "Young gentlemen" was what the midshipmen were called in the old naval service, and Jarvis was the youngest of them all, being just thirteen at the time of the action with the "Vengeance." He was the smallest officer aboard, and his most important duties were those of passing the word from the quarter-deck forward, and taking his station aloft in the maintop, where he was learning the mysteries of the maze of gear which went through the lubber's-hole or belayed in the top. He also stood at quarters with his diminutive sword drawn,—a smaller edition of the lieutenants, who were allowed to wear one epaulette and who could make a louder noise through the speaking-trumpet than Jarvis could hope to for years. Down in the midshipmen's mess, by virtue of his diminutive stature and tender years, he was not much interfered with by Wederstrandt, Henry, Vandyke, and the bigger men. But he fought one or two of the young gentlemen nearer his age, and, though frequently defeated, stood up as strongly as possible for what he deemed his rights. He was a manly little reefer, and up in the maintop, where he was stationed in time of action, the men swore by him. He was sensible enough not to give any orders without the professional opinion of one of the old jackies, who always ventured it with a touch of the cap, a respectful "Sir," and perhaps a half-concealed smile, which was more of interest than amusement. Thirteen was rather a tender age at which to command men of fifty, but the midshipmen of those days were not ordinary boys. They went out from their comfortable homes aboard ships where men were even rougher and less well-disciplined than they are to-day, and they had either to sink or swim. It was Spartan treatment; but a year of it made men and sailors of them or else sent them posting home to their mothers and sisters.

Jarvis loved it, and did his duty like a man. He knew the lead of all the gear on his mast, and kept his few pieces of brass-work aloft shining like new. He kept the rigging in his top, even when there was no occasion for it, coiled down as though for inspection, although nobody but the topmen and yardmen ever had occasion to examine it. He was as active as a monkey, and, scorning the "lubber's-hole," went over the futtock-shrouds as smartly as any of the light-yardmen.

The greatest and probably the only regret of midshipman Jarvis's short life was that he had not joined the great frigate before she met and defeated the "Insurgente" the year before. He wanted to be in a great action. Nothing seemed to make him feel more of a man than when the long 18-pounders were fired in broadside at target-practice. If he had been but a boy, instead of an officer with a gold-laced cap and a dirk and all the dignities pertaining to those habiliments, he would have clapped his hands and shouted for sheer joy. But the eyes of his men were upon him, and so he stood watching the flight of the shots, and biting hard on his lips he kept his composure.

Captain Truxton, ever mindful of his midshipmen, had disposed them in different parts of the ship with regard to their size and usefulness. The older ones had been given gun-divisions, while the youngsters were placed on the fo'c's'le or in the tops, where they might be of assistance, but would more certainly be out of harm's way. Such a thought was not suggested on the "Constellation." If it had been, little Jarvis would probably have resigned immediately, or at the very least have burst into unmanly tears. As it was, he felt that his post aloft was as important as any on the ship, and he promised himself that if another Frenchman was sighted he would stay there whether the mast were up or down.

So, on the 1st of February, 1800, just about a year after the capture of the "Insurgente," while they were bowling along under easy sail, about fifteen miles off Basse Terre, a large sail, which appeared to be a French frigate, was sighted to the southward. Jarvis went aloft two ratlines at a time, his heart bounding with joy at the prospect of the chance of a fight.

On assuring himself that she was a large ship, Captain Truxton immediately set all sail and took a course which soon brought her hull above the horizon and showed the Americans beyond a doubt that she was a ship-of-war of heavier metal than the "Constellation." Nothing daunted, Truxton bore on his course until the gun-streaks of the other vessel could be plainly seen. Instead of showing the same desire to speak, the stranger held on, pointing a little off his course, as though anxious to avoid an encounter.

But the breeze, which had been light, now died away altogether, and the sea became calm. There the two great vessels drifted in sight of each other all night and part of the following day, awaiting the wind which would enable them to close. Jarvis was in a fever of impatience. A half a dozen times he got permission from the officer of the deck, and with a telescope almost as long as himself, clambered up to the main-royal to report. There was but one opinion among the midshipmen who went aloft,—she was a Frenchman. She *could not* be anything else.

About two o'clock in the afternoon of the next day, up to the northward they saw the ripple on the water of the wind they had been waiting for. The sail-loosers flew aloft, and every sail was spread to catch it. Soon the "Constellation" was pushing her way through the water, and the foam was even flying from the wave-tops here and there. The chase had caught the breeze at about the same time, and the Americans could see by the line of white under her bow that she was beginning to leg it at a handsome rate. But the "Constellation" was in excellent condition for a race, and by degrees drew up on the other ship, which as they reached her was seen to lie very low in the water, as though deep-laden. They were sure to discover who she was before nightfall, so Truxton cleared his ship for action. Jarvis went aloft to his top and saw the backstays lashed and the preventer-braces securely hooked and rove. Extra muskets were carried up into his top for the use of the jackies and marines when they should come into close quarters, for then the fire of sharpshooters would be almost as valuable as the shots of the great guns.

Their work had been over an hour and the sun had set in a clear sky before the "Constellation" drew up to gunshot distance. It was moonlight before she came within effective range. The battle-lanterns were lit, and the long row of lights on the Frenchman showed that he, too, was prepared for fight. The sky was clear, and the moon, which was nearly at the full, made the outlines of the vessels perfectly visible to the men at the guns. Jarvis, from his post aloft, could plainly see the lines of heads along the poop, and fancied that he could make out a midshipman almost as young as he, who was clambering about the maintop of the other vessel. He heard the beating of a drum and the sound of cheers as the Frenchmen moved to their quarters.

On the decks below there was not a sound.

Truxton had given his men their orders. There was to be no cheering until there was something to cheer for. They were to await the order to fire until the enemy was close aboard, and then, and not until then, was the broadside to be delivered. The division-officers had gone about quietly repeating these commands to the gun-captains, and there was nothing further to say. Only to wait until the battle began. Jarvis repeated to his topmen, word for word, the instructions he had received, that in their aim particular attention was to be paid to the officers of the enemy.

Soon a gun from the after-battery of the Frenchman was fired. This was followed shortly by all the guns that would bear. Some of the shots crashed into the hull of the "Constellation," and one of them killed several men. The division-officers glanced appealingly to Truxton, in the hope of the order to fire; but he merely held up his hand. Again the broadside came, and men seemed to be falling everywhere. The strain below and aloft was terrific. But the officers stood steadily, with a word of encouragement here and there, and the men did not flinch.

At last the "Constellation" came abreast the after-ports of the Frenchman, and Truxton, throwing her off a little, so that all his

THE "CONSTELLATION" AND THE "VENGEANCE"

broadside would bear in a diagonal direction, loudly shouted the order to fire.

The telling broadside was delivered, and the battle was on in earnest. To those aloft the crash of the long eighteens into the hull of the enemy at every other downward roll of the "Constellation" showed how well the American gunners had learned to shoot, while the short bark of the cannonades and the shrieks in the brief pauses from the decks of the Frenchman told of the terrible effects of the fire among the enemy. The guns of the Frenchman were well served and rapidly fired, but they were aiming on the upward roll of the sea, and their shots went high. Several balls from the smaller pieces had lodged in the foremast and mainmast, and one had struck just below the futtock-band of the maintop, where Jarvis was, and sent the splinters flying up and all about him. Yard-arm to yard-arm they sailed for three long, bloody hours, until the firing of the Frenchman gradually slackened and then stopped almost altogether. The Americans had suffered less on the decks than aloft, and Jarvis's topmen were employed most of the time in splicing and re-reeving gear. The discharge of the "Constellation's" broadside-guns did not diminish for a moment, and so fast was the firing that many of the guns became overheated, and the men had to crawl out of the exposed ports to draw up buckets of water to cool them.

At about midnight Truxton managed to draw ahead of his adversary in the smoke, and, taking a raking position, sent in such a broadside that the Frenchman was silenced completely.

Jarvis and the men in the maintop had little time to use their muskets. Several long shots had struck the mast, and almost every shroud and backstay had been carried away. As the "Constellation" bore down upon her adversary to deal her the death-blow, the mast began swaying frightfully. There was a cry from the men at Jarvis's side, and the marines and topmen began dropping through the lubber's-hole, swinging themselves down the sides of the swaying mast by whatever gear they could lay their hands to.

Jarvis did not move. One of the older seamen took him by the shoulder and urged him to go below. The mast was going, he said, and it meant certain death to stay aloft.

Little Jarvis smiled at him. "This is my post of duty," he replied, "and I am going to stay here until ordered below."

At this moment a terrific crackling was heard, and the old man-o'-warsman went over the edge of the top. All the strain was on one or two of the shrouds, and, just as h reached the deck, with a tremendous crash the great mast went over the side.

Jarvis had kept his promise to stay by his mast whether it was up or down.

The Frenchman, not so badly injured aloft, took advantage of the condition of the "Constellation," and, slowly making sail before the wreck was cleared away, faded into the night. It was afterwards discovered that she was the "Vengeance," of fifty-two guns. She succeeded in reaching Curaçoa in a sinking condition.

When the news of the fight reached home, Congress gave Truxton a medal and a sword, and prize money to the officers and crew.

For little Jarvis, the midshipman, who preferred to die at his post, Congress passed a special resolution, which read:

"*Resolved*, That the conduct of James Jarvis, a midshipman in said frigate, who gloriously preferred certain death to an abandonment of his post, is deserving of the highest praise, and that the loss of so promising an officer is a subject of national regret."

History does not show an instance of nobler self-sacrifice, and no such honor as this special act of Congress was received by a boy before or since.

CUSHING AND THE "ALBEMARLE"

ALTHOUGH the Civil War furnished many instances of conspicuous gallantry, so many that most of them remain to-day comparatively unknown, none was more notable than the torpedo exploit of Lieutenant William Barker Cushing. There have been several similar expeditions in our naval history. Before Tripoli, Richard Somers made the ill-fated attempt with the "Intrepid," and in the war with Spain Richmond Hobson sunk the "Merrimac." There is no question that the personal and sentimental aspects of these three hazardous enterprises are similar. All three men were young, and each one knew that he took his life in his hands. Somers, rather than be captured with his powder, destroyed both his ship and himself. Hobson sunk the "Merrimac," but did not succeed in getting her athwart the channel. Cushing, in a torpedo-launch, went under the guns of the enemy, and escaped both death and imprisonment. On the enemy the moral effect of all three exploits must have been the same. Professionally, Cushing's exploit has just this distinction: he was successful. Like Decatur in the recapture of the "Philadelphia," he carried out in every detail the plans he had made. And upon his success the way was opened for the Union fleet, and the hopes of the Confederates fled, for only two seaports in the South —Charleston and Wilmington—remained open to them.

After the great success of the "Merrimac" in Hampton Roads, the Confederates determined to construct a vessel of similar design for use in the Southern rivers and sounds. Under great difficulties they built the "Albemarle" on the Roanoke River, and carried her into action almost before the last rivet was driven. She was a formidable craft in those days, and the shots from the vessels of the Northern fleet went harmlessly against her iron sides to break and fly into a thousand pieces. On the 5th of May the "Albemarle" had another fight with a larger fleet of Union vessels, which had gathered to hem in and disable her. During the action the "Sassacus" saw an opportunity to ram her, and, going ahead at full speed, struck the ram a terrific blow amidships. The bow of the "Sassacus" was literally torn to pieces by the impact, and the "Albemarle," though heeling over and in danger of sinking for a time, finally righted and pulled out of the action uninjured, but by no means disabled. All of the vessels of the squadron kept up a heavy fire upon her, but she went on to her anchorage up the river, where a few repairs made her as good as ever.

It looked to the Unionists as though the story of the "Merrimac" with the "Congress" and the "Cumberland" was about to be repeated; that the "Albemarle" in course of time would come down at her leisure and destroy all Northern vessels in those waters. To make matters worse, the Unionists learned that another vessel of a similar type was nearly completed, and that the two vessels would attack at the same time,—a combination which, with their consorts, seemed irresistible. Something had to be done if the command of the sounds of the Carolinas was to remain with the navy of the North.

But during the summer of 1864 two steam launches rigged up as torpedo-boats, the invention of Engineer J. L. Long, were fitted out at New York and brought down through the canals to Albemarle Sound. The bows of the boats were cut under and decked over, and the engines were so built that when covered and moving at a low rate of speed they made little or no noise. A spar ten or fifteen feet long, which carried a torpedo and a firing attachment, projected forward over the bow, and a small howitzer was also mounted forward where it would be useful to repel attack.

The government had decided to make a night attempt on the "Albemarle," and the honor of the command of the expedition was bestowed on Lieutenant Cushing, who had half a dozen times before received the thanks of the secretary of the navy for gallantry in action off Cape Fear River.

The expedition was favored by the inactivity of the Confederates. The "Albemarle" lay alongside the dock at Plymouth awaiting the completion of her sister-ship, but this needless delay gave Cushing the opportunity he wanted.

The Confederates were fully aware of the plans of the Unionist's navy, and a thousand soldiers remained to guard the "Albemarle" from land attack as well as to act as sentries for a distance along the river bank. To provide against torpedoes, a line of great cypress logs was boomed off her sides at a distance of twenty to thirty feet, so that it seemed impossible to come within striking distance. Besides this, the smaller guns of the ram were trained up and down the river,—which here was but one hundred and fifty yards wide,—to sweep the entire area over which the attacking party had to pass.

But Cushing, like Decatur, rejoiced at obstacles. He was only twenty-one, but he carried a man's head on his broad shoulders, and the planning of such an expedition down to the smallest detail was a task which he entered into with judgment and enthusiasm, ingredients as rare as they are necessary in such a desperate enterprise.

After a week spent in preparation and experiment, the gunboat "Otsego" brought the launch to the mouth of the river, where Cushing cast off and pointed his bow toward Plymouth, towing a cutter full of armed men, who were to capture, if possible, a Confederate guard,—which had been set in a schooner near the sunken "Southfield,"—to prevent their giving the alarm. But the expedition started badly, for the launch ran aground on a bar. Before Cushing could float her again it was too late to make the attempt. Cushing and his boat's crews then returned to the "Otsego."

The next night was black and squally, with occasional showers of rain. They could only make out the loom of the shore by straining their eyes into the darkness. Cushing was as cool as though taking shore-liberty. As he shook hands with the "Otsego's" officers he paused at the gangway to say, with a laugh,—

"Well, here goes for another stripe or a coffin."

They crept slowly up the river, keeping close to the bank, under the shadow of the reeds and trees. The little engine, covered with tarpaulins, made so little sound that the men in the cutter towing astern could hardly hear it. There was not a sound except the plashing of the gusts of rain and the ripple of the water as the little craft moved steadily on. Cushing knew he must be passing some of the pickets now, so not a word even in whispers was spoken. Every man had his duty and knew when to do it. Acting Ensign William Howarth was aft at Cushing's side. Acting Master's Mate John Woodman, who knew the river, was next to him. The other officers were Acting Master's Mate Thomas S. Gay, Acting Assistant-Paymaster Francis H. Swan, and Acting Third-Assistant-Engineers Charles L. Steever and William Stotesbury.

By half-past two A.M., about a mile below Plymouth, where the "Albemarle" lay, they came upon the submerged "Southfield," and could just make out the lines of the guard-schooner. The machinery of the launch was slowed, almost stopped, for Cushing had decided to get by her if he could without a fight. It was a moment of utmost anxiety, and every man was prepared for the attack. But there was no sign of discovery from the schooners, and in ten minutes the little expedition had passed up the river in safety.

But only the first danger of discovery was over. Between the "Southfield" and Plymouth the bank was guarded by a double line of sentries, and the men in the boats, now moving more quickly, could see the red glare of the smouldering fires reaching all the way to town. As they came to the point beyond which the ram was lying, they found, to their delight, that the fires which should have been brightly burning were smouldering in the rain. There was no sign of a man to be made out anywhere, and Cushing pushed on directly for the "Albemarle," which he could now see plainly as she lay at the wharf, grim and menacing, but without a sign of life.

Suddenly from the shore there came the sharp bark of a dog. To the ears of Cushing and his men in the deep silence and anxiety of the moment it sounded like the report of a Dahlgren. There was a cry from a sentry and a challenge, followed by a musket-shot, and the bullet flew over the boats and struck the water fifty feet away with a vicious *ping* that sounded not less loud than the report itself. There was another challenge, and in a moment the cries came from everywhere. Other dogs began barking, and it seemed as though the whole town was aroused. The sentries on both sides of the stream threw inflammable material on the smouldering fires, and in a moment the river was as bright as day.

Realizing that further concealment was useless, Cushing himself cast off the towline of the cutter, and telling the men in her to row for their lives, gave the engineer the order, "Four bells, ahead full speed," setting the nose of the launch directly for the ram. The sparks flew up from her stack, and the dark water churned up in masses of foam under her stern, as like a sentient thing she leaped forward on her deadly mission. It was then that Cushing discovered for the first time the line of torpedo booms which guarded the ram. In facing the musketry-fire and the great guns of their enormous adversary the task of getting close enough to reach her seemed impossible.

Cushing knew that if he was to get over the log booms he must strike them fair; then perhaps he could slide over within striking distance. He shifted his helm, and making a wide sweep out into the stream, gathered all the headway he could and came down into the very jaws of the great monster. A tremendous volley from the shot-guns and muskets of the sentries greeted them, and Paymaster Swan was wounded. Cushing received a charge of buckshot in the back of his coat and had the sole of his shoe torn off, but these were the only shots which took effect.

Cushing here shouted, in a loud voice, "Leave the ram; we're going to blow you up!" hoping to create a panic. But the Confederates continued firing, and succeeded in getting in two howitzer-shots, one of which felled a man by Cushing's side. At this moment, Gay, up forward in the launch, took careful aim with the howitzer, and sent a charge in the midst of the Confederate crew. Then with a slight jar the sled-like bow rose on the boom. She balanced a second, and slid over within the enclosure, half full of water, but within reaching distance.

One of the great guns of the "Albemarle," a hundred-pounder, protruded from a broadside port directly in front of them, and they could see the gun-crew frantically training the gun and trying to depress the muzzle enough to bear on them. It was a matter of seconds now. Who would fire first? Cushing had lowered the torpedo-spar, and as soon as he had it well under the overhang he detached it, then waited until he heard the torpedo strike the hull, when he pulled the trigger-line. He was not a fraction of a second too soon, for the two concussions were almost simultaneous. There was a muffled roar from below the great vessel, and a column of water shot out from under her as she lifted on the wave. The shock of the hundred-pounder, was terrific, and it seemed as if the frail launch had been blown to pieces. But Cushing had been too quick for them. The charge of canister passed a few feet over their heads and scattered in the river beyond.

The work of the gallant crew was done. Cushing had made a hole in the "Albemarle" large enough to have driven a wagon through. The great wave of the torpedo rose and went completely over the launch, swamping her alongside and throwing her men into the water. All of them got to the booms safely. Here Cushing paused a moment to throw off his outer clothing, while the Confederates on the banks were shouting to the men to surrender. Several of them, being unable to swim, did so; but Cushing, calling to the others to follow him, plunged boldly into the water and struck off down the stream. He was a powerful swimmer, but the night was cold, and he knew that he could not keep up very long. But he swam for half an hour, and he came upon Woodman in the middle of the stream, almost exhausted. Though almost entirely fagged out himself, he tried to help the mate towards the shore. Finding that he was being pulled down and unable to save the other, Cushing struggled on, and reached the shoal water more dead than alive. Here he lay among the reeds until dawn, when he learned from a negro how complete had been his success. At last, after almost twenty-four hours' exposure, he succeeded in finding one of the enemy's deserted picket-skiffs, and managed under cover of the second night to pull off to the Federal "Valley City," which he reached at eleven o'clock at night, and was hauled aboard completely exhausted from his labor and exposure. Only one other of his crew reached a place of safety. Woodman and Samuel Higgins, the fireman, were drowned. The others went ashore and surrendered or were captured.

This service, because of the great benefit to the Union cause and the daring manner in which it had been performed, made Cushing the hero of the year. Congress passed a vote of thanks and promoted him to the rank of lieutenant-commander, which he held until 1872, when he became a commander.

He did not long enjoy his honors, for two

years later he died of brain fever, in Washington, at the age of thirty-two. Had he lived he would have been but fifty-six years of age at the outbreak of the war with Spain, and would have been one of the ranking officers in active service of the new navy.

SOMERS AND THE "INTREPID"

AMONG the young officers of Commodore Preble's squadron before Tripoli there was a tall, dark, melancholy-looking fellow of about twenty-five. His name was Richard Somers and his command was the "Nautilus," a little schooner of twelve guns and a hundred men. He had been with Decatur and Stewart, a junior officer on Commodore Barry's "United States" in the war with Spain, and the friendship formed in those early days had been cemented by a score of thrilling adventures which had drawn them more closely together than brothers. Charles Stewart, before Decatur's promotion to post-captain, had been the second in command to Preble, and his vessel, the "Siren," had taken a prominent part in all the many actions with the Tripolitan forts and gunboats. He was a year or so older than his companions and had drifted a little away from them. But Decatur and Somers were inseparable. Some bond outside of mere professional sympathy and environment existed between them, and there seemed to be no thought of the one that the other did not share. The difference in their temperaments was marked. Decatur was bold, domineering, and impetuous. Somers was quiet, mild, and ever avoided the quarrel which Decatur too often sought. But under the quiet exterior men had found a will like iron and the willingness to dare and do anything that came within the province of his profession. He was thoughtful, but not so quiet that he could not enter into the gayety of the mess; he was mild, but not so mild that he would overlook shortcomings among his men or brook any slight upon his office or his reputation.

In the old days on the "United States" there happened an affair which immediately established his reputation as an officer and a man. At first he was not understood. His brother midshipmen, mistaking the reserve of his manner for weakness, did not hesitate before they had been aboard with him a month to take advantage of him in the steerage and on deck in every possible way. Not only did they slight him, but, after the manner of the cadet midshipman of recent years, they made him the butt of most of their practical jokes below-decks. Somers stood it for a while in silence. He dearly loved peace, and, beyond a good-humored protest, let everything pass for what it was worth. But as the weeks went by and the bantering continued, instead of laughing it off as before, Somers became more and more quiet and self-contained.

Decatur, ever humorous and mischief-making, had himself been one of the worst to chaff his comrade ; but he knew what Somers' silence meant, and he desisted. He had been his school-mate in Philadelphia, and he had seen that ominous quiet before. Decatur would have fought for him to the last drop of his blood, but he felt that his comrade was well able to look out for himself.

Somers went about his duties quietly, never giving a sign that there was anything upon his mind until the day before coming into port, then he went to Decatur, and said,—

"Stephen, to-morrow I want you to go ashore with me, for I am going to meet three men."

The next afternoon a cutter containing Somers, Decatur, and three midshipmen, with their seconds, went ashore and found a secluded spot upon the beach where they would be free from interference. He had challenged all three to fight at the same time and would take them in succession.

In the first two duels Somers received two

shots in the body, the latter one of which caused him to sink upon the sand as though dangerously hurt; but he rallied quickly, and, seeing that the third midshipman was standing waiting to see if the battle could be continued, he tried to struggle to his feet. He found he could not get up, and Decatur offered to take his place and receive the fire of the third midshipman. But Somers, though suffering greatly, was not to be deterred, and bade Decatur prop him up in a sitting posture, in which position he exchanged shots with the third man. Fortunately, none of the injuries resulted fatally, and in a few weeks Somers was on deck again. He went about his duties as quietly as before, but never after that did they call him milksop.

It was Somers who led one division of the gunboats to attack the Tripolitan fleet while Decatur was leading the other. Finding that he could not reach them by the eastern entrance, he sailed into the northern entrance of the harbor and single-handed boldly sent his little vessel into the midst of five of the enemy. His gunboat was smaller than any one of those of his adversaries; but so well was his long gun served and so true was the fire of his musketry that he held them at bay for half an hour, and not one of them succeeded in getting alongside of him to board. They were all bearing straight down upon the rocks, though, and Somers could not spare enough men from the guns to man his sweeps. But Preble, on the "Constitution," saw his danger, and, coming up in time, sent a broadside of grape among the pirates, and they got out their sweeps and retreated, when, in spite of the doggedness of the defence, one united attack would have made the victory theirs. But as they drew off, instead of returning, as Preble wished, to the "Constitution," Somers pursued them until within less than a cable's length of a twelve-gun battery, which had not fired before for fear of damaging the fleeing Tripolitans. When she opened fire at this close range the destruction of Somers's valiant little vessel seemed inevitable. But by a lucky chance a bombard exploded in the battery, blew up the platform, and drove the Tripolitans to cover.

Before they could recover and train their guns, Somers managed to bring his craft out in safety. In a later action, as Somers stood leaning against a flag-staff on his little vessel, a shot came directly for him. The officer saw it in time, and jumped aside to see the spar carried away at just the spot where his head had been. He was spared for more deadly work.

SOMERS AND THE "INTREPID"

While these many attacks were being made upon the gunboats and batteries, the "Intrepid," in which Decatur had recaptured and destroyed the "Philadelphia," was being rapidly prepared as a fire-ship. Their plan was to load her with a hundred barrels of powder in bulk, with bags of grape and solid shot, and under cover of the night explode her in the midst of the Tripolitan war-vessels. Somers, who had been frequently in the harbor of Tripoli and knew its reefs and rocks so that he could readily thread his way through the narrow channels, asked for the opportunity to command this expedition. But Decatur's success in boarding the "Philadelphia" had raised the chivalry of every officer and man in the fleet to a point rarely equalled in our own history, and Somers, while he did not begrudge Decatur his two epaulettes, was filled with the passion to do a deed as great, if not greater. They had been rivals since youth, and he felt that now was the opportunity to attempt a great deed for his country, though he and every man in the fleet knew that the chances of coming out alive were but one in a hundred. Somers went to Commodore Preble and urged his knowledge of the harbor as his chief claim to the service. It was an honor that a half-dozen other men sought, and not until the old commodore had weighed the chances fully did he at last agree to let Somers go. But, before consenting, Preble repeatedly warned the young officer of the desperate character of the work, and told him that on account of the Napoleonic wars the Tripolitans were short of ammunition, and that so much powder must not fall into the hands of the enemy. But Somers needed no warning. A day or two afterwards, when the preparations were nearly completed, Preble and some other officers were trying a fuse in the cabin of the "Constitution." One of the officers, watch in hand, ventured the opinion that it burned too long and might enable the enemy to put it out before it exploded the magazine. Hearing this, Somers said, quietly,—

"I ask for no fuse at all."

He was more gentle than ever in those last few days, and as he and Decatur leaned over the hammock-nettings of "Old Ironsides," looking towards the line of white where the sea was breaking over the outer roofs, the melancholy look seemed to deepen and the far-away expression in his eyes was of another world. Decatur knew that rather than give up his ship and his powder, Somers would blow the ship and himself to eternity.

When volunteers were called for, the desperateness of the enterprise was fully explained; but the crew of the "Nautilus," Somers's own vessel stepped forward to a man. He selected four,—James Simms, Thomas Tompline, James Harris, and William Keith. From the "Constitution" he took William Harrison, Robert Clark, Hugh McCormick, Jacob Williams, Peter Penner, and Isaac Downes. Midshipman Henry Wadsworth (an uncle of the poet Longfellow) was chosen as second in command. Midshipman Joseph Israel, having vainly pleaded with Somers to be allowed to go, at the last moment smuggled himself aboard the "Intrepid," and when discovered Somers had not the heart to send him back.

Decatur and Stewart went aboard the "Nautilus" on the evening that the attempt had been planned. The three had been so closely united all their lives that Stewart and Decatur felt the seriousness of the moment. Even professionally the attempt seemed almost foolhardy, for several Tripolitan vessels had come to anchor just within the entrance, and to pass them even at night seemed an impossibility. Somers felt a premonition of his impending catastrophe, for just as they were about to return to their own vessels he took a ring from his finger and, breaking it into three pieces, gave each of them a part, retaining the third for himself.

As soon as the night fell the "Intrepid" cast off her lines and went slowly up towards the harbor. The "Argus," the "Vixen," and the "Nautilus" followed her, while shortly afterwards Stewart on the "Siren" became so anxious that he followed, too. A haze that had come up when the sun went down hung heavily over the water, and soon the lines of the fire-ship became a mere gray blur against the dark coast-line beyond. The excitement upon the guard-ships now became intense, and both officers and men climbed the rigging and leaned out in the chains in the hope of being able to follow the movements of the ketch. Midshipman Ridgley, on the "Nautilus," by the aid of a powerful night-glass aloft, managed to follow her until she got well within the harbor, and then she vanished. The suspense soon became almost unbearable, for not a shot had been fired and not a sound came from the direction in which she had gone. At about nine o'clock a half-dozen cannon-shots could be plainly heard, and even the knowledge that she had been discovered and was being fired on was a relief from the awful silence.

At about ten o'clock Stewart was standing at the gangway of the "Siren," with Lieutenant Carrol, when the latter, craning his neck out into the night, suddenly exclaimed,—

"Look! See the light!"

Stewart saw away up the harbor a speck of light, as if from a lantern, which moved rapidly, as though it were being carried by some one running along a deck. Then it paused and disappeared from view. In a second a tremendous flame shot up hundreds of feet into the air, and the glare of it was so intense that it seemed close aboard. The flash and shock were so stupendous that the guard-ships, though far out to sea, trembled and shivered like the men who watched and were blinded. The sound of the explosion which followed seemed to shake sea and sky. It was like a hundred thunder-claps, and they could hear the echoes of it go rolling down across the water until it was swallowed up in the silence of the night.

That was all. The officers and the men looked at one another in mute horror. Could anything have lived in the area of that dreadful explosion? The tension upon the men of the little fleet was almost at the breaking point. Every eye was strained towards the harbor and every ear caught eagerly at the faintest sound. Officers and men frequently asked one another the question, "Have you heard anything yet?" with always the same reply.

The vessels beat to and fro between the harbor-entrances, firing rockets and guns for the guidance of possible fugitives. And the doleful sound of that gun made the silences the more depressing. All night long did the fleet keep vigil, but not a shot, a voice, or even a splash came from the harbor.

With the first streaks of dawn the Americans were aloft with their glasses. On the rocks at the northern entrance, through which the "Intrepid" had passed, they saw a mast and fragments of vessels. When the mist cleared they saw that one of the enemy's largest gunboats had disappeared and two others were so badly shattered that they lay upon the shore for repairs.

The details of the occurrence were never actually known, but it is thought that Somers, being laid aboard by three gunboats before actually in the midst of the shipping, and feeling himself overpowered, fired his magazine and destroyed himself and his own men in his avowed purpose not to be taken by the enemy.

Thus died Richard Somers, Henry Wadsworth, the midshipman, Joseph Israel, and ten American seamen, whose names have been inscribed on the navy's roll of fame. Nothing can dim the honor of a man who dies willingly for his country.

THE PASSING OF THE OLD NAVY

OLD SALTS AND NEW SAILORS

SINCE ballad-mongering began, the sea and the men who go down to it in ships have been a fruitful theme; and the conventional song-singing, horn-piping tar of the chanteys is a creature of fancy, pure and simple.

Jack is as honest as any man. Aboard ship he goes about his duties willingly, a creature of habit and environment, with a goodly respect for his "old man" and the articles of war. Ashore he is an innocent,—a brand for the burning, with a half-month's pay and a devouring thirst.

Sailor-men all over the world are the same, and will be throughout all time, except in so far as their life is improved by new conditions. Though Jack aboard ship is the greatest grumbler in the world, ashore he loves all the world, and likes to be taken for the sailor of the songs. In a week he will spend the earnings of many months, and go back aboard ship, sadder, perhaps, but never a wiser man.

He seldom makes resolutions, however, and so, when anchor takes ground again, his money leaves him with the same merry clink as before. Though a Bohemian and a nomad, he does not silently steal away, like the Arab. His goings, like his comings, are accompanied with much carousing and song-singing; and the sweetheart he leaves gets to know that wiving is not for him. With anchor atrip and helm alee, Jack mourns not, no matter whither bound.

The improved conditions on the modern men-of-war have changed things for him somewhat, and, though still impregnated with old ideas, Jack is more temperate, more foresighted, and more self-reliant than he once was. His lapses of discipline and his falls from grace are less frequent than of yore, for he has to keep an eye to windward if he expects to win any of the benefits that are generously held out to the hard-working, sober, and deserving.

But the bitterness of the old days is barely disguised in the jollity of the chanteys. However we take it, the sea-life is a hardship the like of which no land-lubber knows. Stories of the trials of the merchant service come to him now and then and open his eyes to the real conditions of the service.

THE PASSING OF THE OLD NAVY

Men are greater brutes at sea than ashore. The one-man power, absolute, supreme in the old days, when all license was free and monarchies trod heavily on weak necks, led men to deeds of violence and death, whenever violence and death seemed the easiest methods of enforcing discipline. Men were knocked down hatchways, struck with belaying-pins, made to toe the seam on small provocation or on no provocation at all. The old-fashioned sea-yarns of Captain Marryat ring true as far as they go, but they do not go far enough.

In England the great frigates were generally both under-manned and badly victualled, and the cruises were long and sickening. The practice of medicine had not reached the dignity of the precise science it is to-day, and the surgeon's appliances were rude and roughly manipulated. Anæsthetics were unknown, and after the battles, the slaughter in which was sometimes terrific, many a poor chap was sent to his last account by unwise amputation or bad treatment after the operation.

The water frequently became putrid, and this, with the lack of fresh vegetables and the over use of pork, brought on the disease called scurvy, which oftentimes wiped out entire crews in its deadly ravages. Every year thousands of men were carried off by it. A far greater number died from the effects of scurvy than from the enemy's fire. Lieutenant Kelly says that during the Seven Years' War but one thousand five hundred and twelve seamen and marines were killed, but one hundred and thirty-three thousand died of disease or were reported missing. Not until the beginning of this century was this dreadful evil ameliorated.

The evils of impressment and the work of the crimp and his gang—so infamous in England—had no great vogue here, for the reason that, during our wars of 1776 and 1812, the good seamen—coasters and fishermen, who had suffered most from the Lion—were only too anxious to find a berth on an American man-of-war, where they could do yeoman's service against their cruel oppressor.

"Keel-hauling" and the "cat" were relics of the barbarism of the old English navy. Keel-hauling was an extreme punishment, for the unfortunate rarely, if ever, survived the ordeal. In brief, it consisted in sending the poor sailor-man on a voyage of discovery along the keel of the vessel. Trussed like a fowl, he was lowered over the bows of the ship and hauled along underneath her until he made his appearance at the stern, half or wholly drowned, and terribly cut all over the body by the sea-growth on the ship's bottom. He bled in every part from the cuts of the barnacles; but "this was considered rather advantageous than otherwise, as the loss of blood restored the patient, if he were not quite drowned, and the consequence was that one out of three, it is said, have been known to recover from their enforced submarine excursion."

Think of it! Recovery was not anticipated, but if the victim got well, the officer in command made no objection! Beside the brutality of these old English navy bullies a barbarous Hottentot chief would be an angel of mercy.

Flogging and the use of the cat were abolished in the American navy in 1805. This law meant the use of the cat-o'-nine tails as a regular punishment, but did not prohibit blows to enforce immediate obedience. Before that time it was a common practice for the punishment of minor offences as well as the more serious ones.

Flogging in the old days was an affair of much ceremony on board men-of-war. The entire ship's company was piped on deck for the punishment, and the culprit, stripped to the waist, was brought to the mast. The boatswain's mate, cat in hand, stood by the side of a suspended grating in the gangway, and the captain, officer of the deck, and the surgeon took their posts opposite him. The offence and the sentence were then read, and the stripes were administered on the bare back of the offender, a petty officer standing by to count the blows of the lash, while the doctor, with his hand on the victim's pulse, was ready to give the danger signal when absolutely necessary.

The men bore it in different ways. The old hands gritted their teeth philosophically, but the younger men frequently shrieked in their agony as the pitiless lash wound itself around the tender flesh, raising at first livid red welts, and afterwards lacerating the flesh and tearing the back into bloody seams.

The effect upon the lookers-on was varied. The younger officers, newly come from well-ordered English homes, frequently fainted at the sight. But the horror of the spectacle soon died away, and before many weeks had passed, with hardened looks, they stood on the quarter-deck and watched the performance amusedly. Soon the spectacle got to be a part of their life, and the jokes were many and the laughter loud at the victim's expense. The greater the suffering the more pleasurable the excitement.

Many yarns are spun of Jack's tricks to avoid the lash or to reduce to a minimum the pain of the blows. Sometimes the men had their flogging served to them regularly, but in small doses. To these the punishment lost its rigor. For the boatswain's mate not infrequently disguised the force of his blows, which came lightly enough, though the victim bawled vigorously to keep up the deception, and in the "three- and four-dozen" cases he sometimes tempered his blows to the physical condition of the sufferers, who otherwise would have swooned with the pain.

One Jacky, who thought himself wiser than his fellows, in order to escape his next dozen, had a picture of a crucifix tattooed over the whole surface of his back, and under it a legend, which intimated that blows upon the image would be a sacrilege. When next he was brought before the mast he showed it to the boatswain and his captain. The captain, a crusty barnacle of the old harsh school, smiled grimly.

"Don't desecrate the picture, bos'n," he said; "we will respect this man's religious scruples. You may put on his shirt," he said, chuckling to himself, "but remove his trousers, bos'n, and give him a dozen extra. And lay them on religiously, bos'n."

All this was in the older days, and it was never so bad in the American as in the English navy. The middle period of the American navy, from before the Civil War to the age of iron and steel cruisers, presents an entirely different aspect in some ways.

Illegal punishments were still inflicted, for there were always then, as now, a certain percentage of ruffians forward who were amenable to no discipline, and could be managed only by meeting them with their own weapons. The "spread-eagle" and the ride on the "gray mare" were still resorted to to compel obedience.

They "spread-eagled" a man by tricing him up inside the rigging, taut lines holding his arms and legs outstretched to the farthest shrouds, a bight of rope passed around his body preventing too great a strain. He was gagged, and so he could not answer back.

The "gray mare" on which the obstreperous were forced to gallop was the spanker-boom—the long spar that extends far over the water at the ship's stern. By casting loose the sheets, the boom rolled briskly from side to side, and the lonely horseman was forced in this perilous position to hold himself by digging his nails into the soft wood or swinging to any of

THE SMOKING HOUR

the gear that flew into his reach. At best it was not a safe saddle, and a rough sea made it worse than a bucking broncho.

Paul Jones had a neat way of disciplining his midshipmen aloft. He would go to the rail himself, and casting loose the halyards, let the yard go down with a run, to the young gentleman's great discomfiture.

But the life of the old salt was not all bitterness. It was not all shore-leave, but there was skittles now and then for the deserving and good-conduct men. Jack's pleasures were simple, as they are to-day. There was never a crew that did not have its merry chanter and its flute, fiddle, or guitar, or the twice-told tale of the ship's Methuselah to entertain the dog-watches of the evening or the smoking-hour and make a break in the dreary monotony of routine.

On public holidays, when everything was snug at sea or in port, a glorious skylark was the order of the afternoon. At the call of the bos'n's mate, "All hands frolic," rigorous discipline was suspended, and the men turned to with a will to make the day one to be talked about. Mast-head-races, potato- and sack-races, climbing the greased pole, and rough horse-play and man-handling filled the afternoon until hammocks were piped down and the watch was set. Purses from the wardroom and prizes of rum and tobacco—luxuries dear to Jack's heart—were the incentives to vigorous athletics and rough buffoonery. The rigging was filled from netting to top with the rough, jesting figures, and cheer upon cheer and laugh upon laugh greeted a successful bout or fortunate sally.

Jack is a child at the best of times and at the worst, and he takes his pleasures with the zest of a boy of seven, laughing and making merry until he falls to the deck from very weariness. And woe be at these merry times to the shipmate who has no sense of humor. His day is a hideous one, for he is hazed and bullied until he is forced in self-defence to seek the seclusion granted by the nethermost part of the hold. A practical joker always, when discipline is lax, Jack's boisterous humor knows no restraint.

The ceremony of "crossing the line," the boarding of the ship by Neptune and his court, seems almost as old as ships, and is honored even to-day, when much of the romantic seems to have passed out of sea-life. It is the time when the deep-sea sailor has the better of his cousin of the coasts. Every man who crossed the equator for the first time had to pay due honor to the god of the seas.

NEPTUNE COMES ABOARD

They exacted it, too, among the whalers when they crossed the Arctic Circle.

The wardroom usually bought off in rum, money, or tobacco, but forward it was the roughest kind of rough man-handling; and the victims were happy indeed when they got their deep-water credentials. The details of procedure in this remarkable rite differed somewhat on different ships, but the essential elements of play and torture were the same in all cases.

The day before the line was to be reached both wardroom and forecastle would receive a manifesto setting forth the intention of the god of the seas to honor their poor craft and ordering all those who had not paid tribute to him to gather forward to greet him as he came over the side. At the hour appointed there was a commotion forward, and a figure, wearing a pasteboard crown that surmounted a genial red face adorned with oakum whiskers, made its appearance over the windward nettings and proclaimed its identity as Neptune. Behind him was a motley crew in costumes of any kind and all kinds—or no kind—who had girded itself for this ungentle art of bull-baiting. The deep-water men intended to have an ample return for what they themselves had suffered, not many years back, when they had rounded the Horn or Cape of Good Hope.

The unfortunates, stripped to the waist, were brought forward, one by one, to be put through their paces. After a mock trial by the jury of buffoons, the king ordered their punishment meted out in doses proportioned directly to the popularity of the victims as shipmates. The old long boat, with thwarts removed and a canvas lining, served as a ducking-pond. After vigorous applications, of "slush,"—which is another name for ship's grease,—or perhaps a toss in a hammock or a blanket, they were pitched backward into the pool and given a thorough sousing, emerging somewhat the worse for wear, but happy that the business was finally done for good and all.

To-day the roughest sort of bullying no longer takes place, and much of the romance seems to have passed out of the custom.

The punishments, too, have lost their severity. The "gray mare" swings to an empty saddle, the "spread eagle" is a thing of the past, and the "cat" is looked upon as a relic of barbarism. Things are not yet Pinafore-like, but the cursing and man-handling are not what they used to be. There are a few of the old-timers who still believe the "cat" a necessary evil, and would like to see an occasional "spread eagle," but the more moderate punishments of to-day have proved, save in a few hardened cases, that much may be done if the morale of the service is high.

The fact of the matter is, that the standard of the man behind the gun has kept up with the marvellous advance of the ships and the ordnance. To-day, the naval service of the United States is worthy of any seaman's metal. As a mode of living, sea-faring on American men-of-war attracts as many good men as any other trade. Machinists, electricians, carpenters, gunners, and sail-makers, all have the chance of a good living, with prizes for the honest and industrious.

The seaman himself, in times of peace, may rise by faithful service to a competency and a retiring pension more generous than that of any other nation in the world. The discipline is the discipline of right relations between superior and inferior men of sense, and the articles of war govern as rigorously the cabin as the forecastle. Republican principles are carried out, as far as they are compatible with perfect subordination, and there exists no feeling between the parts of the ship, except in extraordinary instances, but wholesome respect and convention. There is little tyranny on the one side or insubordination on the other.

The training of the young officer of the old navy was the training of the larger school of the world. "Least squares" and "ballistics" were not for him. He could muster a watch, bend and set a stun'sail, work out a traverse, and pass a weather-earing; but he toyed not with the higher mathematics, like the machine-made "young gentleman" of to-day. What he knew of navigation he had picked hap-hazard, as best he might.

At the age of twelve his career usually opened briskly in the thunder of a hurricane or the slaughter of a battle, under conditions trying to the souls of bronzed, bearded men. Physical and even mental training of a certain kind he had, but the intellectual development of modern days was missing. The American officer of the days before the Naval Academy was founded was the result of rough conditions that Nature shaped to her own ends with the only tools she had. Though these "boys" had not the beautiful theory of the thing, they had its practice, and no better seamen ever lived.

At the beginning of the century, the crusty Preble, commodore of the blockading fleet before Tripoli, was sent a consignment of these "boys" to aid him in his work. The names of the "boys" were Decatur, Stewart,

Macdonough, Lawrence, and Perry. Excepting Decatur, who was twenty-six, there was not one who was over twenty-four, and two or three of them were under twenty. The commodore grew red in the face and swore mighty oaths when he thought of the things he had to accomplish with the youngsters under his command. But he found before long that though youth might be inconvenient, it could not be considered as a reproach in their case.

Decatur, with a volunteer crew, went under the guns at Tripoli, captured and blew up the "Philadelphia" in a way that paled all deeds of gallantry done before or since. The dreamy Somers went in with a fire-ship and destroyed both the shipping and himself. In the hand-to-hand fights on the gunboats, Lawrence, young Bainbridge, Stewart, and the others fought and defeated the best hand-to-hand fighters of the Mediterranean. The Dey of Algiers, when Decatur came before him to make terms of peace, stroked his black beard and looked at the young hero curiously. "Why," he said, "do they send over these young boys to treat with the older Powers?"

When the war was over, Preble no longer grew red in the face or swore. He loved his school-boys, and walked his quarter-deck with them arm-in-arm. And they loved him for his very crustiness, for they knew that back of it all was a man.

These youthful heroes were not the only ones. Young Farragut, an infant of twelve years, with an old "Shoot-if-you're-lucky," quelled a promising mutiny. At eighteen Bainbridge did the same. Farragut, at thirteen, was recommended for promotion to a lieutenancy he was too young to take. Perry was about thirty when he won the victory of Erie.

A youngster's character bears a certain definite relation to the times he lives in. Skies blue and breezes light, he shapes his life's course with no cares but the betterment of his mental condition. Baffling winds create the sailor, and storm and stress bring out his greater capabilities. The Spanish war has proved that heroes only slumber, and that the young gentleman with the finely-tempered mind of an Annapolis training is capable of the great things his father did.

The blue-jacket of to-day has plenty of hard work to do, but he is as comfortable as good food and sleeping accommodations, regular habits, and good government can make him. As a class, the United States Jacky is more contented, perhaps, than any other man of similar conditions. Unlike the soldier, he does not even have to rough it very much, for wherever he goes he takes his house with him.

Jacky sleeps in a hammock strung upon hooks to the beams of the deck above him. When he turns out, he lashes his hammock with its lashing, and stores it in the nettings,—the troughs for the purpose at the sides of the ship,—where it must stay until night. If Jack wants to sleep in the meanwhile, he chooses the softest spot he can find on a steel-clad deck; and he can sleep there, too, in the broad glare of daylight, a hundred feet passing him, and the usual run of ship's calls and noises droning in his ears.

Jacky's food is provided by the government, while his superior of the wardroom has to pay his own mess-bill. He is allowed, in addition to his pay, the sum of nine dollars per month, and this must purchase everything, except such luxuries as he may choose to buy from his pay. The ship's paymaster is allowed a certain amount of money to furnish the supplies, and between him and the ship's cook the problem is settled. At the end of the month, if the amount served out is in excess of the computation for rations, the brunt falls upon the "Jack-of-the-Dust,"—the assistant to the paymaster's yeoman,—who has the work of accurately measuring the rations which are given to the cook of the ship.

The ship's cook receives from the government from twenty-five to thirty-five dollars a month, according to the size of the ship, and, in addition, certain money perquisites from the different messes, which gives him a fair average. He has complete charge of the ship's galley and the cooks of the messes, and must be able to concoct a dainty French dish for the wardroom as well as the usual "salt horse" or "dog" for the Jacky.

"Salt horse" is the sea-name for pork. "Dog" is soaked hardtack, mixed with molasses and fried; and, though it is not pleasant twenty-nine days out of the month, it is healthful, and tastes good to a hard-working sailor with the salt of the sea producing a splendid appetite.

The mess-tables hang by iron supports to the beams of the deck above, and when the mess has been served and eaten,—as only Jack knows how to eat,—they are triced up into their places, and all is cleaned and made ship-shape in the twinkling of an eye. A half-hour is allowed for dinner, and this time is kept sacred for Jack's use. A red pennant flies from the yard-arm, that all may know that the sailor-man is eating and must not be disturbed by any importunate or curious callers.

In the dog-watches of the evening, after supper, from six to eight P.M., the blue-jacket is given his leisure. It is then that pipes are smoked, vigilance relaxed, boxing and wrestling bouts are in order, and Jacky settles down for his rest after the day of labor. From somewhere down on the gun-deck comes the tinkle of a guitar or banjo, and a tuneful, manly voice sings the songs of France or Spain, and, better still, of beloved America, for the shipmates.

The sailor of to-day is also a soldier. Back in the days of Henry the Eighth, when England first had a navy, the sailors only worked the ships. The fighting was done by the soldiers. Later, when the ships were armed with many guns and carried a greater spread of canvas, there was no space for great companies of soldiers, and the sailors became gunners as well. A few soldiers there were, but these did only sentry duty and performed the duties of the ships's police. As such they were cordially hated by the jackies.

This antipathy has come down through the ages to the present day, and marines are still looked on by the sailor-men as land-lubbers and Johnnies—sea-people who have no mission upon the earth save to do all the eating and very little of the rough work.

The new navy has done much to change this feeling. The mission of the marine is now a definite one. Always used as a sharpshooter, he now mans the rapid-fire batteries, and even guns of a larger caliber. He has done his work well, and the affair at Guantanamo has caused the sneer to fade from the lip of the American sailor-man. Two of the ablest captains of our navy, always the deadliest opponents of the marine corps, upon assuming their latest commands, applied immediately for the largest complement of marines that they could get.

Any ship, old or new, is as frail as the crew that mans it. The strength of any vessel varies directly with its discipline and personnel. Hull, Jones, Decatur, Bainbridge, and Stewart, in the old days, knew with some accuracy the forces they had to reckon with. Their guns were of simple contrivance, and their men knew them as well as they knew how to reef a topsail or smartly pass a weather-earing. They feared nothing so long as they were confident of their captain. New and mysterious contrivances for death-dealing were unknown to them, and hence the morale of the old sea-battles was the morale only of strength and discipline.

There were no uncertain factors to reckon with, save the weight of metal and the comparative training of the gun-crews.

To-day the unknown plays a large part in warfare. Intricate appliances, mysterious inventions, new types of torpedo-boats, and submarine vessels form a new element to contend against and have a personal moral influence upon the discipline of crews. To combat this new element of the unknown and uncertain has required sailors and men of a different stripe from the old. Where, in the old days, ignorance and all its accompanying evils held sway over the mind of poor Jack, and made him a prey to superstition and imagination, to-day, by dint of careful training of brain as well as body, he has become a thinking creature of power and force of mind. He knows in a general way the working of the great mechanical contrivances; and in the fights that are to come, as well as those that have been, he will show that the metal the American Jacky is made of rings true and stands well the trial by fire.

PIKE AND CUTLASS

THE OLD SHIPS AND THE NEW

With much hitching of trousers and shifting of quid, the old longshoreman will tell you that sea-life isn't at all what it once was.

He will gaze out to sea, where the great iron machines are plying back and forth, and a reminiscent sparkle will come into his eyes as he turns to his lobster-pots and tells you how it was in the good days of clippers and sailing-frigates, when sailor-men were sailor-men and not boiler-room swabs, machine-made and steam-soaked. He will also yarn, with much d—ning of his eyes (and yours), of how fair it was in the deck-watches of the "Saucy Sally" barque, with everything drawing alow and aloft, grog and 'baccy a-plenty, and never a care but the hurry to spend the voyage-money. And not till he's mumbled all his discontent will he haul his sheets and give you right-of-way.

He forgets, sheer hulk that he is, that he's been in dry-dock a generation or more and that swift-moving Time has loosed his gear and dimmed his binnacle-lights. Despite his ancient croaking, tricks at the wheel are to-day as ably kept, eyes as sharp as his still peer into the dimness over the forecastle, and the sea-lead takes as long a heave as in the early sixties, when he hauled up to New York with a thousand dollars in prize-money and a heart full for the business of spending it. It has always been so. There has never been an age that has not had its carper to tell you of the wonders that once were.

Yet it was truly beautiful. With the tide on the ebb and the wind a-piping free, never was a fairer sight than the Atlantic clipper as she picked her speedy way through the shipping to the harbor's mouth; and nothing so stately as the gallant frigate in her wake, with all sail set to ga'n's'ls, her topsails bellying grandly to the quartering breeze, which whipped the filmy wave-tops against her broad bows, under which the yellow curl lapped merrily its greeting. The harbor clear and the capes abeam, aloft flew the nimble sail-loosers. The royals and the stu'n-sails flapped to the freshening wind, sheets went home with a run, and the yards flew to their blocks.

Then, her departure taken, like a gull she sped blithely on her course. The rays of the afternoon sun gilded her snowy canvases until she looked a thing of air and fairyland, not of reality. On she flew, her tall spars dipping grandly to the swells—a stately farewell courtesy to the clipper, hull down to leeward. On the decks the boatswain piped his cheerful note, and everything came ship-shape and

Bristol-fashion for the cruise. The running-gear was neatly coiled for running, the guns secured for sea, and the watches told off. The officer of the deck walked to and fro, singing softly to himself, casting now and then a careful eye aloft to the weather-leeches, which quivered like an aspen as the helmsman, leaning to the slant of the deck, kept her well up to her work.

And yet the poetry has not gone out of it all. The poetry of the sailing-frigate was lyric. That of the steel battle-ship is Homeric.

Nothing save a war of the elements has the power of a battle-ship in action. Ten thousand tons of steel,—a mighty fortress churning speedily through the water fills the spirit with wonder at the works of man and makes any engine for his destruction a possibility. Away down below the water-line a score or more of furnaces, white-heated, roar furiously under the forced draught, and the monster engines move their ponderous arms majestically, and in rhythm and harmony mask their awful strength. Before the furnace-doors, blackened, half-naked stokers move, silhouetted against the crimson glare, like grim phantoms of the Shades. The iron uprights and tools are hot to their touch, the purple gases hiss

SEA MONSTERS IN ACTION

and sputter in their very faces, yet still they toil on, gasping for breath, their tongues cleaving to their mouths, and their wet bodies steaming in the heat of it.

The deck above gives no sign of the struggle below. Where, in the old days, the sonorous trumpet rang out and the spar-deck was alive with the watch who hurried to the pin-rail at the frequent call, now all is quiet. Here and there bright work is polished, or a lookout passes a cheery call, but nothing save the man at the wheel and the officer of the watch shows the actual working of the ship.

Seamanship, in the sense of sail-handling, is a thing of the past. Though there is no officer in the navy who could not in an emergency handle a square-rigger with the science of an old sea-captain, the man on the bridge has now come to be first a tactician and after that a master of steam and electricity.

In the sea-battles of 1812 the captain was here, there, and everywhere in the thickest of the fight, inspiring by his personal magnetism the men at the guns. He was the soul of his ship. To-day the sea-battle is a one-man battle. The captain is still the heart and soul of the ship, but his ends are accomplished in a less personal way. His men need not see him. By the touch of a finger he can perform every action necessary to carry his ship to victory. He can see everything, do everything, and make his presence everywhere felt by the mere operation of a set of electrical instruments in front of him.

The intricacies of his position are, in a way, increased. He may lose a boiler, split a crank, or break an electrical connection, but the beautiful subtleties of old-fashioned seamanship have no place whatever on the modern warship.

Let it not be understood that the handling of the great ocean fortress of to-day may be mastered by any save a craftsman of the art. With plenty of sea-room and a keen watch alow and aloft the trick is a simple one, for the monster is only a speck in the infinity of sea and sky, and there is never a fear save for a blow, or a ship, or a shore. But in close manœuvre, or in harbor, the problem is different. Ten thousand tons of bulk cannot be turned and twisted on the heel with the swish and toss of the wieldy clipper. Observant transpontine voyagers, who have watched the gigantic liner warped out from her pier into a swift tide-way with a leeward ebb, will tell you what a complicated and difficult thing it seems to be.

The captain of the battle-ship must be all

that the merchant captain is, and more besides. Mooring and slipping moorings should be an open book to the naval officer, but his higher studies, the deeper intricacies of the science of war, are mysteries for the merchant captain. All of it is seamanship, of course. But to-day it is the seamanship of the bridled elements, where strength is met by strength and steam and iron make wind and wave as nothing.

The perfection of the seamanship of the past was not in strength, but in yielding, and the saltiest sea-captain was he who cajoled both ship and sea to his bidding. The wind and waves, they say, are always on the side of the ablest navigators, but it was rather a mysterious and subtle knowledge of the habits and humors of God's sea and sky, and a sympathy born of constant communion, which made both ship and captain a part of the elements about them, and turned them into servants, and not masters.

The naval captains of 1812 had learned this freemasonry of sea and sky, and one incident —a typical one—will show it as no mere words can do. Its characteristics are Yankee pluck and old-fashioned Yankee seamanship.

The frigate "Constitution"—of glorious memory—in 1812 gave the British squadron which surrounded her startling proof of the niceties of Yankee seamanship. There never has been a race for such a stake, and never will be. Had "Old Ironsides" been captured, there is no telling what would have been the deadly effect on the American fortunes It was the race for the life of a nation.

The "Constitution" was the country's hope and pride, and Captain Hull knew it. He felt that "Old Ironsides" could never fail to do the work required of her. So for four days and nights the old man towed her along, the British frigates just out of range, until he showed clean heels to the entire squadron. The ingenuity and deft manœuvring of the chase has no parallel in the history of this or any other country in the world.

With hardly a catspaw of wind, Hull drifted into sight of the British fleet off the Jersey coast. Before he knew it, they brought the wind up with them, and his position was desperate. There were four frigates and a ship-of-the-line spread out in a way to take advantage of any breath of air. Hull called away his boats, and running lines to them, sent them ahead to tow her as best they might. The British did still better, for they concentrated the boats of the squadron on two ships, and gained rapidly on the American. Hull cut ports over the stern, and ran two 18-pounders out of his cabin windows, where he began a continuous fire on the enemy. The British ships shifted their helms and took up positions on the quarters of the frigate, unable to approach too closely with their boats for fear of the "Constitution's" stern-guns, which dropped their hurtling shot under their very bows.

The desperate game had only begun. Hull, finding that he had but one hundred and fifty feet of water under him, decided to kedge her along. In a few minutes the largest boat was rowing away ahead with a small anchor on board, stretching out half a mile of cable. The anchor dropped, the men hauled in roundly and walked away with the line at a smart pace. It was heart-breaking work, but the speed of the ship was trebled. By the time the vessel was warped up to the first anchor another one was ready for her, and she clawed still further out of the enemy's reach. The British did not at first discover the magic headway of the American, and not for some time did they attempt to follow suit.

Then a breeze came up. Hull hauled his yards to it, picked up his boats without slacking sail, and went ahead. But hardly were the sails drawing when the wind died away again. One of the ships came into range, and there was nothing for it but to go back to the kedging. Three times did this occur, the captain, with his eye on the dog-vane, jockeying her along as a skipper would his racing-yacht. The men had now been at their quarters for thirty-six hours without rest or sleep. But at the order they dropped into the boats again, ready for anything.

Another breeze sprang up now and held for two hours. Like logs the sailor-men tumbled over on the decks, nearly dead for lack of sleep. On the afternoon of the third day of the chase the "Constitution" lost the wind and the enemy kept it. Back again to kedging they went, weary and sick at heart.

But relief was in sight. A great cloud hove up on the southeastern horizon, and the black squall that followed was a Godsend to the "Constitution" and her weary crew. Hull knew the Englishmen would not like the looks of the squall. No more did he. But he kept his boats at the towing, nevertheless.

He stationed his men at the halyards and down-hauls, and had everything in hand for the shock. He calmly watched the on-coming line of froth, growing whiter every minute, while his officers came to him and begged him to take in his sail. But wait he did until the first breath stirred his royals. Then the shrill pipe of the boatswain called the boats alongside of the "Constitution."

They were not a moment too soon. As the men were hooking the tackles the blast struck the ship. Over she heeled, almost on her beam ends, the boats tossed up like feather-weights. The yards came down with a rush, and the sails flew up to the quarter-blocks, though the wind seemed likely to blow them out of the bolt-ropes. She righted herself in a moment, though, and so cleverly had Hull watched his time that not a boat was lost.

Among the enemy all was disorganization. Every sail was furled, and some of their boats went adrift. Then, as the friendly rain and mist came down, the wily Yankee spread his sails—not even furled—and sailed away on an easy bowline at nine knots an hour.

The race was won. Before the Englishmen could recover, Hull managed by wetting his sails to make them hold the wind, and soon the enemy was but a blur on his western horizon. Then the British gave it up.

The superiority of Yankee seamanship was never more marked than in this chase. The British had the wind, the advantage of position, the force, and lacked only the wonderful skill and indomitable perseverance of the American, who, with everything against him, never for a moment despaired of pulling gallant "Old Ironsides" out of the reach of his slow-moving enemy.

The difficult manœuvre of picking up his boats without backing a yard or easing a sheet he repeated again and again, to the wonderment of his adversaries, whose attempts in this direction failed every time they tried it in a smart breeze. Hull's tactics at the coming of the squall were hazardous, and under any other circumstances would have been suicidal. For a skipper to have his boats two cable-lengths away from his ship, with his royals flapping to the first shock of a squall, is bad seamanship. But if tackles are hooked and men are safe aboard there is no marine feat like it.

The naval history of this country is full of such instances. Captain Charles Stewart, on the same ship, did a wonderful thing. In his fight with the "Cyane" and the "Levant" he delivered a broadside from both batteries at the same time. Then, shifting his helm under cover of the smoke, he backed his topsails and drew out sternward from the enemy's fire, taking a new position, and delivering another broadside, which brought about their surrender.

The war-ship of fifty years ago was as different from the battle-ship of to-day as a caravel from a torpedo-boat. With half the length and a third the tonnage, the old "ship-of-the-line" had three times as many men as the modern sea-fighter. Yet, with a thousand men aboard, she had work for them all. More than two acres of canvas were to be handled, and over a hundred guns were to be served, loaded, and fired. A thousand pieces of running-gear were to be rove and manned. The huge top-sails, weighing, with their yards, many tons, needed on their halyards half a hundred men. Great anchors were to be broken from their sandy holds, and the capstan-bars, double-banked, hove around to the sound of the merry chantey and deep-voiced trumpet. Homeward bound, the business of anchor-hoisting turned into a mad scene, and many a rude jest and hoarse song turned the crowded fo'c's'le into a carnival of jollity.

In matters of routine and training the crews of the American frigates differed little from those of England. The sailor-men of the United States, though newer to the work of navigating the big ships, were smart seamen, and could cross or bring down their light yards, send down their masts, or clear for action with the oldest and very best of England's men-o'-war's-men.

The ships themselves differed little in general construction. During the war of 1812, of large frigates we possessed but the "Constitution," the "President," and the "Constellation." Though built upon models patterned after the accepted standards of the period, they were somewhat smaller than the British vessels and usually carried a lighter armament. Their unbroken list of victories during the war with England is remarkable when one considers what the young nation was contending against, both at home and abroad, and how little aid Congress had given the infant navy.

It seems really wonderful how a large body of men, numbering from three hundred to six hundred, and later a thousand or more, could find comfort and a home from one year's end to another in a space only two hundred feet long and fifty feet wide.

But Jack is nowhere so comfortable as aboard ship. He is used to prescribed limits, and crawls into his hammock at night happy that the space is no greater. There is a companionship, he thinks, in close quarters, and he likes them.

In the old ships it was a matter of great importance to provide comfortable quarters for the great crews they were obliged to carry. In England, during the first years of the century, the complement of a "Seventy-four" was five hundred and ninety, and even six hundred and forty men. Hammocks seem to have been used during the reign of Queen Elizabeth, when they were called "nets," probably because they were made of rope-yarn.

The officers were then, as now, given the after part of the ship. A wooden bulkhead separated the cabins of the officers from the main-decks, where the men lived, though when the ship was cleared for action the bulkheads were taken down and all movable property both of officer and man was taken below-deck.

This gave a clean sweep of the deck from bow to stern. The steerage had from two to six broadside-guns in it, and even the captain had to live with a couple of brass stern-chasers and a broadsider or two.

The grandest line-of-battle-ship ever built for this country was the old "Pennsylvania." She was made of wood throughout, two hundred and twenty feet long and fifty-eight feet beam, with a draft of twenty-five feet of water and thirty-five hundred tons displacement,—just one-third of that of the modern "Iowa." Eleven hundred men could swing their hammocks on her wide decks, where no modern gun-carriages or steel compartments broke the long sweep from the cabin forward. Her sides were of oak, with a thickness of eighteen inches at the upper gunports and thirty-two inches at the water-line, almost heavy enough at long range to resist the shot of a modern rifle. Her sixteen inches were proof against her own fire at a mile. On her three fighting-decks she carried sixteen 8-inch guns, the heaviest they had in those days, and one hundred and four 32-pounders. Her mainmast was over two hundred feet long, and with all sail set she could leg it at twelve knots an hour.

But compare her with the modern "Indiana." The "Pennsylvania" weighed less than the armor of the "Indiana" alone. The "Indiana" has but sixteen guns, against one hundred and twenty on the "Pennsylvania;" but that broadside can send two tons of tempered steel at a single discharge. The old 8-inch guns of the "Pennsylvania" could send a shell through fifteen inches of oak at a distance of a mile—the equivalent of half an inch of steel.

The range of a modern rifle is from five to twelve miles; the penetration is almost anything you please in the way of steel armor. The "Pennsylvania's" shells at point-blank range would hardly make a perceptible dent in the "Indiana's" steel armor, and the old cast-iron shot would roll harmlessly down the new ship's sides. But one explosive shell from the "Indiana" would go through the "Pennsylvania" from stem to stern, and would splinter and burn her beyond repair.

The "Pennsylvania" cost the government, in 1837, nearly seven hundred thousand dollars; a fabulous sum for a battle-ship in those days. The "Indiana" cost three millions and a half,—only two hundred and fifty thousand dollars less than the sum paid for that vast territory bought from Napoleon, and known as the "Louisiana Purchase," and about half the sum paid for the acquisition of Alaska from Russia.

The statistics are interesting. According to official authority, in putting this vessel together seven hundred tons of rivets alone were used. About four hundred plans were made for the hull and about two hundred and fifty plans and drawings were made for the engines. These would take a force of one hundred men a year to complete.

The engines and machinery alone weigh about nine hundred tons. The smoke-stacks are about sixteen feet in diameter. Each of the main engines is so enormous that under the great frames, in the economy of space and construction, are two smaller engines, the sole mission of which is to start the big ones. There are about sixty-six separate engines for various purposes. The condensing-tubes, placed end to end, would cover a distance of twelve miles. Thirty tons of water fill her boilers, which would stand a pressure of one hundred and sixty pounds to the square inch. Three dynamos provide the electricity,—a plant which would light a town of five thousand inhabitants. There are twenty-one complete sets of speaking-tubes and twenty-four telephone stations.

The two great turrets are clad with nineteen inches of toughened steel. In each of these turrets are two 13-inch guns. Each of these guns is about fifty feet long and weighs sixty-one tons. There are eight 8-inch guns on the superstructure, in sets of twos, and amidships on the main-deck are four 6-inch rifles. In ten minutes, firing each 13-inch gun once in two minutes, and using all the other guns at their full power, the "Indiana" could fire about sixty tons of death-dealing metal.

The millennium has not yet been reached, but such awful force makes universal peace a possibility. What the immediate future holds forth in naval architecture and gunnery is a matter which excites some curiosity, for it almost seems as though perfection, according to the standard of the end of the century, has been reached. And yet we already know of certain changes, improvements, and inventions, the direct outcome of the Spanish war, which are to be made on the vessels now contracted for, which affect importantly the government of the ship; and so it may be that the next twenty years will show as great an evolution as have the two decades just past.

But whatever the future may bring, it has been a marvellous and momentous change from the old navy to the new. Since the "Monitor"-"Merrimac" fight no country has been quicker to profit by the lessons of the victory of iron over wood and steel over iron than the United States.

But the navy that is, however glorious its achievements, can never dim the glory of the navy that was, though sailor-men, old and new, know that in a test of ship and ship, and man and man, the flag of this country will continue to fly triumphant.

FARRAGUT IN MOBILE BAY

IT was Friday, the 5th of August, 1864. The first violet streaks of dawn stole through the purple clouds that the wind had tossed up during the night. Admiral Farragut sat in his cabin, quietly sipping his tea, his fleet-captain, Drayton, by his side. Through the open ports they could see the dim masses of the ships of the fleet as, lashed two and two, they stretched in a long line to seaward. The wind no longer blew, and the shrill pipes and the creaking of the blocks as the light yards came down echoed clearly across the silent water.

"How is the wind, Drayton?" said the admiral, at last.

Drayton walked to the port.

"About west-sou'west, sir, I should say."

The admiral smiled.

"A good omen. Our smoke will blow over their batteries."

He raised his cup, drained it, and set it back on its saucer. Then he rose to his feet and walked slowly up and down the cabin, looking first at his watch and then out through the starboard gallery, where the fleet lay. He turned, his genial face all aglow in the cool light of the morning, and reached to the table for his side-arms.

The moment had arrived.

"Well, Drayton," he said, "we might as well get under weigh."

Drayton knew, and Farragut knew, that the momentous day before them would decide the fate of the West Gulf and of the nation in the South. It was the supreme moment in the admiral's career. But as he clasped his sword-belt his hands were as firm as though on inspection.

With a cheery "Aye, aye, sir," Drayton went out of the door and up the companion, and soon the deck above resounded with the nimble feet as the men sprang joyfully to quarters. Old Knowles, the quartermaster, deftly sent his little ball of bunting, ready for an hour, to the yard-arm, and in a moment the row of multi-colored flags, tipped with the glow of the brightened east, fluttered proudly out into the morning breeze.

Then the bright answering pennants flew up from all the vessels of the fleet, and the black smoke poured from their dusky funnels as the white water churned up behind them on their way into line.

The admiral, on the quarter-deck, glass in hand, saw the black turrets of the monitors, with their grim, shiny muzzles, drift slowly inland towards the batteries, not a ripple showing behind them as they moved on their deadly mission towards the frowning battlements of Fort Morgan. Ahead of the "Hartford" was the broad stern of the "Brooklyn," as she churned her way slowly onward, her smoke drifting in great clouds over her starboard bow towards the water-batteries. Beside the admiral, one hand on the rail, was Drayton, cool as though on a practice drill, and as he looked over the swarthy backs that shone bare in the morning sun he knew well that the flagship would give a good account of herself.

Behind him stood Watson, Gates, McKinley, and Brownell, watching the progress of the monitors. The calmness of the scene was sublime. Only an occasional order to the tacklemen, given in a quiet voice by the gun-captains, showed the deadly work ahead.

As the "Hartford" drew into range, the admiral walked over to the main rigging and clambered up into the shrouds; and his men below him at the batteries lovingly watched their "old man" as step by step he mounted to get a clearer view. They knew him for a gallant old sea-dog. They had seen him steam past the batteries at Vicksburg and Port Hudson, and they smiled at his sternness at the capture of New Orleans, for they loved him. But at Mobile they learned that he feared nothing above the ocean or under it, if it stood in the way of the cause of his country. At this point Farragut stood a few feet above Jouett, on the wheel-house of the "Metacomet" alongside, and could hail the top above him, where Freeman, his trusty pilot, gave him his soundings and bearings.

At length the battle opened. A great puff of white smoke rolled along the water from the turret of the "Tecumseh," and a yellow cloud of dust above the water-batteries marked where the shot had struck. Fort Morgan immediately replied, and, as the gunners got the range, the angry splash of the shots as they skipped across the water came clearly to the crew of the "Hartford," who stood at their guns silent and motionless. As the shots rained about them and great white splinters were torn from the nettings and flew across the decks, they only looked up at their admiral, who, leaning slightly forward, was slowly scanning the breastworks. In his face there was no impatience, no irritation, no sign of anxiety, and while he could calmly wait, they could. The courage of the leader was reflected in his men. It was the very perfection of human discipline.

Would the order to fire never come? Already a fragment of shell had struck a gun-captain in the breast, and they saw him carried past them, moaning piteously. A shot had struck the foremast, and a jagged splinter from the mainmast flew up and lodged in the rigging below where the admiral stood. They saw him take the glass from his eyes, and, turning towards Captain Drayton, hold up his hand.

The guns, already trained, belched forth their iron greeting to the gunboats, and the battle was on in earnest. Calm before, the men were calmer now, and they went about their work as though at target practice. The powder-boys flew like sprites, and the gunners sponged and loaded with rapidity. It was as if each gun and its crew were parts of one mechanism.

"Steady, boys, steady. Left tackle a little. So! so!"

And then came another broadside, followed by an eager cheer as the enemy were driven away from their water-battery.

As the smoke from the broadsides increased and obscured his view, the admiral, ratline by ratline, ascended the rigging until he found

THE ADMIRAL LASHED TO THE RIGGING

himself partly above the futtock bands and holding on to the futtock shrouds. The watchful eye of Drayton saw him perched high up, all unconscious of himself, thinking only of the great movements about him. A shock, and he would be thrown into the sea. The captain gave an order to Knowles, the quartermaster, who lay aloft briskly with a piece of lead-line. The admiral did not even see him, and only when Knowles passed the line around him did Farragut take his glasses down. "Never mind," said he, with a smile, "I'm all right." But the quartermaster lashed him, nevertheless, and lay below.

Then from his lofty position the admiral saw a magnificent but terrible thing. The monitor "Tecumseh" was up well with the fort, and drawing slowly on, when, without a warning, a great column of water shot up under her starboard bow. She heeled over to port and went down with every soul on board. She had struck a torpedo. Captain Craven, in his eagerness to engage the "Tennessee" in battle, had passed to the west of the fatal buoy.

This disaster was not immediately realized by the men. Some supposed the "Tennessee" had been sunk, and cheer after cheer was taken up and echoed along the line.

But the admiral knew the danger that was

coming. His anxiety was not decreased when the "Brooklyn," just ahead of him, suddenly stopped. The frown on his brows deepened, and loudly he hailed his pilot, Freeman, in the top, a few feet above him,—

"What's the matter with the Brooklyn?" he shouted. "She must have plenty of water there."

Freeman's head appeared promptly at the lubber's hole.

"Plenty and to spare, admiral," he answered.

Then the admiral knew. Captain Alden had seen the "Tecumseh" go down, and the heavy line of torpedoes across the channel made him pause. The backing screw churned up the water, and the "Hartford" every moment was bearing down on her. The vessels in the rear, pressing on those in the van, created a terrible confusion, and in the uncertainty the batteries of Farragut's ships ceased fire, while the whole of Mobile Point was a living flame. Disaster was imminent.

But not a second did Farragut pause. A harsh voice from the "Hartford" broke the brief but ominous silence.

"What's the trouble?"

Then Alden's voice from the "Brooklyn" answered,—

"Torpedoes."

"Damn the torpedoes!" shouted the admiral. "Four bells. Captain Drayton, go ahead. Jouett, full speed."

And the "Hartford" dashed forward, passed the "Brooklyn," and assumed the head of the column.

Over the line of mines they flew at full speed, and the men below could hear them as they scraped along the hull. It was the one way out of the difficulty, and a second's hesitation would have closed even this escape from a frightful calamity. The admiral looked astern at the manœuvring of his vessels with a smile of satisfaction. It was a magnificent sight. At first they appeared to be fouling each other in dire confusion, at the mercy of the guns which still belched forth a merciless fire. But as the "Hartford" dashed forward, one by one, as if by magic, they took their places. And he knew a grand tactical movement had been accomplished.

Nor did he forget the poor men of the "Tecumseh," struggling in the water where their ship had gone down, but, going down the rigging, ordered Jouett to lower a boat immediately and pick up the survivors.

The "Hartford" was nearly a mile ahead before the line could be straightened, and single-handed she fought the batteries and the gunboats, making straight for Buchanan's invincible ram, the "Tennessee." Amid the fire of shot and bursting shell the admiral walked calmly back to his quarter-deck, giving a word of advice here and an order there. But soon the other vessels were able to pour in a storm of shot and shell that completely silenced the batteries.

One by one he saw the gunboats sink, until only the "Tennessee" had to be accounted for. The admiral tried to ram her, and the solid shot of his broadsides rolled down her iron sides; but she slipped away, pouring in a terrific fire at close range. She riddled the "Brooklyn," "Richmond, and "Monongahela," all three of which dashed at her, bows on, at fearful speed. The admiral again struck her a fearful blow, but apparently with no effect whatever.

The ram had one great advantage: she was surrounded by enemies and could fire continually, while the Union vessels had to use the utmost care not to fire into or collide with one another. An accident of this kind now happened to Farragut's ship. The "Hartford" and the "Lackawanna" were both making at full speed for the ram. The "Hartford" had the better position; and the "Lackawanna," sheering off to avoid another ship, ran into the quarter of the flagship, just where the admiral was standing, cutting her down nearly to the water's edge. The shock of the impact nearly took him off his feet, but in a moment he was climbing over the side to see what damage had been done.

His crew thought he was looking out for himself. Immediately there was a cry, "Get the admiral out of the ship." The whole thought of his crew, unmindful of themselves, was to get him to a place of safety. It was a mere sudden impulse. But Farragut was not the man to look to himself. Having satisfied himself that the "Hartford" could last, he again gave the order, "Full speed," and set his prow again for the "Tennessee."

But in the meanwhile the monitors had been hammering away at her with their heavy shot. Her rudder and smoke-stack were shot away, and her shutters jammed, and as the "Hartford" bore down upon her for the third time she showed her white flag and surrendered.

The "Hartford" was greatly cut up,—twenty-five killed and twenty-eight wounded,—but the admiral had not a scratch to show for his deadly encounters. He came on deck just as the poor fellows who had been killed were being carefully laid out on the port side of the quarter-deck.

"It was a great victory, Drayton," said he, sadly, "but——"

And the men saw him turn aside, tears coursing down his cheeks.

In truth, "there is nothing half so melancholy as a battle lost, except a battle won."

AT THE NAVAL ACADEMY

IN times like those we have but recently passed through, when the theories and studies of thirty years are being put to tests of fire and the sword, it is interesting to turn for a moment to our naval school at Annapolis, where the officers who planned our campaigns, directed our battles and our blockades, and commanded our ships were first trained to the serious business of war. Though the years which have passed since 1861 have made changes in the personnel system and appearance of the Naval Academy, the city of Annapolis itself is the same sleepy, careless, happy-go-lucky town of earlier days.

Once a year, and only once, it rouses itself from its lethargy and assumes an air of gayety and importance which it may not even have shown when it earned for itself the title of "The Gayest Colonial Capital." During the latter part of May and the first of June each train that pulls into the ramshackle station bears a load of pretty young women,—sisters, cousins, sweethearts,—who come for the two-weeks' exercises, when the naval cadets are graduated, and for the June ball. It has been so since the founding of the Naval Academy, and will be so as long as youngsters in brass buttons are brought up to be professional heroes.

In the old colonial days Annapolis was rich. There was an English governor, and grouped about him were some of the oldest English families. In the middle of the eighteenth century Annapolis had become refined, gay, elegant, and even dissipated.

Not only was Annapolis in these old days the most lucrative place in the colonies for the practice of law, but it was the birthplace of such lawyers as Daniel Dulaney, William Pinckney, Charles Carroll, and Reverdy Johnson. In those days, too, after the Revolution, Charles Carroll of Carrollton, the richest man in America, was one of the citizens. To-day, while the descendants of some of these families are still in possession of the homes of their forefathers, the seat of power and money of Maryland has changed to the commercial capital, Baltimore. The centre of social gayety, therefore, is to be found in the Naval Academy.

The social feature of the life of the cadet must not be underestimated. The youngsters who present themselves as candidates for admission, appointed politically, come from all parts of the country, and represent every shade of opinion and training in the United States. They are a smaller image of the large mass of our people. The problem of bringing these different natures into accord with the conditions which they must face is no easy one; and the weeding-out process, which immediately begins, is conducted by the superintendent—usually a captain in the navy—and the officers under his command, under rules which have been adopted after sixty years of previous administrations.

There is an indefinable something in the organization of the place that makes an indelible impression upon the mind of the candidate, and as he enters upon his duties it does not take long to discover whether he is mentally and personally fitted for the long task before him. It was said in the old days that a seaman was born and not made. But modern warfare has so changed the conditions that, while the officers of the navy must always command men and have the instincts of the sailor, high mental attainments are also the requisite, and those instincts can be formed by experience and association.

The course, then, in brief, is the training of the mind and the body, the school of the soldier and sailor, and the school of the gentleman. Here, then, is where the social influences of the Naval Academy are felt. Politics, like misfortune, makes strange bedfellows, and the scion of your Eastern banker may soon find himself detailed as the room-mate of the most impecunious and unpretentious of Uncle Sam's younger sons. It is the democracy of military training, in which every man's standing is governed alone by his professional qualifications. Money or position can in no way affect his life. His rise or fall depends entirely upon his own worth.

To the young man fortunate enough to secure an early appointment from his representative in Congress, his new home, in the month of May, presents every attraction. From the moment he passes the gate, passes the marine guards, his eye meets the beautifully kept lawns of the campus and drill-ground, sweeping gradually down to the sea-wall on the north and east sides, where the Severn River flows, stretching out to the blue waters of Chesapeake Bay, only three miles from old Fort Severn. To the left, as he enters, are the New Quarters and hospital. To the right, the sacred precincts of "Lovers' Lane," into which he cannot go, under pain of displeasure of his upper classmen, until he has passed through the first, or "plebe," year, and this rule is stringent.

To pass the examinations successfully the candidate must be physically sound, and must have a knowledge of arithmetic, geography, United States history, reading, writing, spelling, English grammar, and the first principles of algebra. The number of appointees is limited by law to one naval cadet for every member or delegate of the House of Representatives, one for the District of Columbia, and ten at-large; the District of Columbia and the at-large appointments being made by the President. The course of the naval cadets is six years,—four years at the Naval Academy and two years at sea,—at the expiration of which time the cadet returns for the final graduation.

The fourth-class man who enters in May has a certain advantage over the September appointee, for he has the advantage of four months of practical instruction, which hardens his muscles and gets his mind into excellent shape for the harder work of the year. Having passed his examinations, the youngster goes to the office of the superintendent, where he takes the oath of allegiance which binds him to serve in the United States navy eight years, including his time of probation at the Naval Academy, unless sooner discharged. He deposits a sum of money for his books, and such other amount as may be necessary for his outfit, and is put to no further expense.

His pay is five hundred dollars a year while at the Naval Academy, but, while he acknowledges its receipt to the paymaster by signing the pay-roll, he is furnished with only sufficient pocket-money to get along on. This sum of money is microscopic, and is usually spent as soon as received. Having procured his outfit from the storekeeper, he reports on board the "Santee." The "Santee" is one of the old sailing-frigates in the navy, and has for years been anchored at the naval dock as quarters for cadets during the summer time and for practical instruction in the drill of the old Dahlgrens. Here, too, is where the fractious cadets are placed in durance.

Until within a very few years the new fourth-class men were sent upon the summer cruise of cadets, first on the "Dale," then on the "Constellation" and the "Monongahela." But by a change in the curriculum the May appointees in the fourth class do not take the summer cruise. The "Monongahela," one summer, carried the line division of the first class, the second class, and the third class. Before this change the life of the "plebe" on the summer cruise was not a bed of roses.

The cadets of the third class, until recently "plebes" themselves, were prepared to wreak upon their juniors all of the pent-up exuberance of the previous year.

Hazing, in the old sense, has died away, and even the "running" of ten years ago has been reduced to a minimum through the efforts of Captains Ramsey, Sampson, and Phythian; but the "plebe" was made to step around in a very lively manner, and to do most of the hauling on the heavy gear, while the third-class men did the complaining. On the "Monongahela" the first, second, and third classes are now, as in the old days, considered as sailors, although a number of the blue-jackets are retained on the vessel. The cadets do their share of the work, and perform all the duties of men-of-war's-men except scrubbing, holy-stoning, and cleaning brass-work. The lower-class men are divided into watches with the regular blue-jackets, side by side with whom they assist in performing all the evolutions in working the ship.

The cruise which follows is usually a pleasant one. There is a lot of hard work to do, and in a short while the hands and muscles get hard, the white suits conveniently tarry, and the skins of the youngsters as brown as leather. But the life has its compensations, for at Fortress Monroe they get into their uniforms again and go ashore to the dances given there at the time of their arrival and departure.

Meanwhile the engineer division of the first class is off on a cruise to visit the various navy-yards and docks of the Atlantic coast. Their course of instruction differs from that of the cadets on the "Monongahela," and they are shown the practical side of engineering work on sea-going ships. Away down below the water-line of their vessel, in the stoke-hole, engine-room, or boiler-room, covered with grease or coal-dust, they do all the work of oilers, engineers, stokers, and mechanics, so as to be able to know accurately all the duties of those men, and to be able to command them in the years to come.

In October the study-term begins, and the cadets are then given their quarters for winter. Most of them are in the building known as the New Quarters, while the others, cadet officers of the first class, are placed in the Old Quarters. The subtle distinction in the titles of these two sets of buildings is hardly appreciated at the Naval Academy, since they have both been built for thirty or forty years, and are in a frightful state of dilapidation. Two cadets of the same class are quartered in each room, and the discipline of household, as well as of person, begins immediately. Each room is plainly furnished, and contains two beds, two wardrobes, two looking-glasses, two iron wash-stands, a common table, and a broom. The charge of the room is taken by each cadet every other week, and this cadet is responsible for its general order and cleanliness. If the officer in charge should happen to inspect the quarters in his absence, and find anything contrary to regulations, the cadet in charge is the one who is reported at the next morning's formation, although his room-mate may have been the delinquent.

Throughout the year the reveille sounds at six o'clock. At a quarter to seven is morning formation, roll-call, and inspection. The ranks are opened, and the keen-eyed officer in charge, followed by the cadet officer-of-the-day and his ominous scratch-pad, with keen eyes looks for grease spots, specks of dust on blouses, tumbled hair, or unblackened boots. After breakfast the sick-call is sounded, and cadets who are ill, or who think they are, report to the hospital. At eight o'clock the study begins, and lasts until half-past twelve. The cadets of each class are divided into sections of from six to a dozen each, and at the bugle-call are formed by sections and marched to their recitation-rooms for study. The morning is divided into two parts, and each part is divided into two periods, one for study and one for recitation.

Briefly, the course of instruction is as follows: Fourth class, first year: algebra, geometry, English, history of Greece and Rome, French, naval history of the United States, Spanish. Third class, second year: descriptive geometry, trigonometry, the Constitution of the United States, analytical geometry, mechanical drawing, physics, and chemistry. Second class, third year: seamanship, principles of mechanism, differential calculus, integral calculus, physics, chemistry, mechanical drawing, and navigation. First class, line division: seamanship and naval tactics, ordnance and gunnery, theory and practice of navigation, hydrographic surveying, least squares, applied mechanics, naval construction, ballistics, armor, and torpedoes. The engineer division has marine engines, boilers, machinery designing, mechanics, and naval construction.

The first part of the course, it will be seen, deals with the simpler branches of study. The plan is not to burden the mind of the cadet with unnecessary knowledge, yet every branch which will directly, or even indirectly, contribute to his ultimate efficiency has its place in the curriculum. The end—the making of a thoroughly trained seaman—is kept constantly in view. The simpler studies train the mind of the cadet to the technical work which follows in the third and fourth years, and in those two years he gets his principal technical and practical training. Each one of the departments in which he studies has a head, usually a naval officer above the rank of lieutenant-commander. All of these heads of departments, with the superintendent and commandant of cadets, who is also head of the Department of Discipline, form the Academic Board. The afternoon classes begin at two and last till four, after which comes the afternoon drill, which lasts until 5.30 and completes the daily duties.

It does not seem with all this work as though the cadet had very much time to himself, but the cadet is not unhappy. Wednesday and Saturday afternoons are given over as recreation-hours, and football and baseball with neighboring college teams bring crowds of visitors into the Academy. The band plays upon the lawn, and the pathways are filled with fair visitors, who walk with their respective heroes along the shady lanes. Saturday night, too, during the winter, hops are given, sometimes by officers and sometimes by cadets, and a gymnastic entertainment once a year gives the cadets the opportunity to show their prowess in boxing, fencing, and work on the gymnastic paraphernalia.

Towards the end of May the annual exercises begin. The examinations finished, the arrival of the Board of Visitors is announced by the booming of cannons from the sea-wall. The cadets receive them on dress-parade, and the work of showing their progress during the year is at once begun. The Board of Visitors go out on one of the government tugs into Chesapeake Bay, and there they see the upper-class men tack, wear-ship, box, haul, and perform all the evolutions in a seamanlike manner on the old "Monongahela." Light yards are swung across with the precision of old men-of-war's-men; sails are reefed, furled, or set in an incomparably short space of time; and the cadets are down from aloft for their target practice. The target is towed out by a launch, anchored, and gun by gun, battery by battery, division by division, or by broadside, the cadets hammer away at it as though it were the vessel of a hostile power, more often than not blowing it entirely to pieces.

Back again at the yard, they go through with their drill as infantry or artillery; and last, but not least, comes the drill by companies for the honor of bearing the Naval Academy flag during the coming year. The judges in this

REEFING TOP-SAILS

competition are usually army officers, and every movement is carefully watched and marked. The captain of each company, before going to this drill, selects its sponsor,—a very pretty girl, who, the drill over, presents the flag to the victorious company amid loud cheers from the whole battalion.

The exercises are over. The cadet of the first class is now ready to be graduated. Companies are formed up in hollow square, and the secretary of the navy in the centre, with a pleasant word to each, presents the diplomas to the graduates amid cheers from the companies. As quickly as he can the first-class man goes to his quarters and shifts into his new uniform, and comes back to the campus for the congratulations of his friends. That night the June ball takes place, and the graduate bids farewell to his old associations and goes out into the world.

Few articles that have been written about the Naval Academy have given anything of the personal side of the life of the cadet,—the side of his life that is an escape-valve from books and drills. There was a time, years ago, when smoking was permitted by the superintendent, and this is how the privilege was granted: One night, in January, 1879, an alarm of fire was sounded just before ten o'clock. The cadets, then nearly ready for turning in, appeared in all sorts of costumes, but reported promptly in the hall. When the battalion was assembled at fire-quarters, word passed that there was a fire in the city and they were expected to aid.

With a cheer the cadets dashed to the engines, and, in spite of the cold and their scanty costumes, rushed out to the State-House circle, where seven or eight buildings were all ablaze.

It was found that the hydrants could not supply enough water, so the cadet officers immediately took charge and ran a line of hose to the river. Four houses were already past help, but attention was immediately directed towards saving the others.

In order to save three buildings it was found necessary to pull one of these burning structures down. A heavy chain was passed through the doors and one of the windows, which was manned by the cadets and townsfolk, and the building was in a short time demolished. In some unaccountable way, after part of the building had been pulled down, the chain was unshackled, and the townsfolk, who were now manning it, shot half-way up the street. So the cadets, in spite of their hard work, could always find time for skylarking. One officer, who was not very much liked, received the full force of the hose, which was in charge of two cadets, directly under the chin. Of course, apologies were in order, but the officer had to go home. At four o'clock in the morning the cadets, wet and tired out, returned to their quarters.

The next day they found that it was generally considered that they had not only saved the buildings but the greater part of the business portion of the town, as the wind had shifted, and the part of the town towards the harbor would have been completely destroyed. At formation the order of the superintendent was read. It said that, "Whereas, the cadets had shown great bravery in the performance of their duty the night before, and had conducted themselves in a creditable manner, the superintendent desired to express his appreciation and grant to them the privilege of using tobacco." Ten minutes after breakfast there was not a man in the battalion of nearly four hundred who was not puffing away furiously on pipe, cigar, or cigarette, although not an ounce of tobacco had been drawn from the stock of the storekeeper. Whence it came is a mystery.

The privilege was taken away in 1881; and though to-day there is no smoking allowed, and smoking is considered one of the most serious offences, yet it is safe to say that in many a secret nook this contraband is safely hid from the eye of the officer in charge. In the old days, after taps, or lights out, poker-parties were the order of the night. The windows and transoms were covered with blankets, and every ray was hidden from the eye of the zealous officer and watchman. But to-day the discipline is different, and the cadet, to pass the rigorous mental examination, has no time to transgress the written and unwritten law.

There are, of course, many criticisms from various quarters as to the methods of instruction at the Naval Academy, but it is not desirable to make rapid changes, in spite of new conditions, in a course that has proved successful for many years. It is asked that if cadets are to man steamships without sails, what is the use of educating them to officer sailing-vessels? What was the necessity of building the "Bancroft," if she was not to be used for the practice-cruises of the cadets? Why has it been proposed to build wooden vessels for their instruction? The superintendent of the Naval Academy, Captain Cooper, Secretary Herbert, and Secretary Long have contended that officer-like qualities can best be attained by experience in sailing-vessels. They believe that intrepidity and alertness come from the old school of sailing-ships.

On the other hand, many of the older officers believe that there is too much book-learning at the Academy and too little practical instruction; but most of them are willing to admit that the naval officer of to-day must be a scientific man to properly meet requirements of modern ships, and that he cannot acquit himself properly unless he has a complete theoretical training. It is certain that the cadet graduated now from the Naval Academy is thoroughly trained in his profession. He has never yet been shown deficient in knowledge of any duty which he has been called upon to perform, nor incapable of mastering the intricate parts of modern ships. Considering the age at which he leaves the Academy, he is better educated in his profession than the college graduate, and is also trained in those qualities for command which make the American naval service what it is to-day. He goes forth thoroughly equipped for his life-work.

OUR NATION'S NEW HEROES

THE great General Grant, when a cadet, went through his course at West Point with one foot out of the Academy and the other in. So curiously deficient was he in all the arts and sciences which theory insists must go to make the perfect soldier that he was always in the "Immortals."

"Immortals" is the name of the section at the foot of the class, admission to whose profane cult means small marks and the possible privilege of resigning at the end of the half-year. Immortals is a neat contraction of "Les Immortals,"—that is, lazy mortals. Immortal Grant became, but not in the way the academic reports of the time would have indicated.

This has proved true again and again among the gradutes of the Naval Academy, as well as those of West Point. Though the "child is father to the man" in general tendencies and character, it does not follow that mere mental attainments are an indication of great genius in the practical operation of the great military professions. Works of the brain and works of the body and spirit are two things; and though the finely-ordered mind controls to a degree both body and spirit, no such mechanism can ever accomplish great deeds in which heart and spirit are needed, though it may plan the details with a nicety to challenge criticism. A combination of all these qualities is rare, for the bookworm is seldom an enthusiast on any subject which gets very far away from his theories.

DOES SCHOLARSHIP COUNT IN WAR?

The Spanish war has shown that it is not always the men who stand at the heads of their classes who lead in the more practical duties of ship and camp. Admiral Sampson, one of the greatest thinkers and most profound students in the navy, as a boy and as a man always led in everything he undertook; but, on the other hand, Hobson, though one of the leaders of his class at Annapolis, was demure and retiring, hardly the man one would select to lead a forlorn hope into the jaws of death.

One may go through the list man for man, and find as many backward in their studies as those who have carved high niches for themselves in the Academy records.

No proposition could cover the situation in a general way, for, after all, the men we have heard from were perhaps only lucky,—lucky in being chosen as the instruments of the result. There are hundreds—thousands—of officers in the service, some brilliant, some wise, some brave, some strong, as good as they, who have lacked only opportunity. The singling out of any names for special mention seems an injustice to them,—"the heroes of the heart."

TAYLOR AND EVANS AS SCHOOL-MATES

Forty years ago Harry Taylor and Bob Evans were boys together in Washington. They were school-mates and chums, fighting each other's battles and longing for the day when they would be old enough to go to the Naval Academy and fight for their country. They were both lively, active lads, Taylor perhaps the quieter of the two.

As their characters developed, Taylor became more of a student than Evans, and that became the distinguishing feature of their entire careers. While Captain Taylor has been the student of books, Captain Evans is known throughout the navy as a student of men and a "man's man" in the best sense of the term. The friendship of youth continued without break throughout their young manhood and prime. The bond was strengthened when

Evans, at the close of the Civil War, married his chum's sister.

They were both in the famous three-year class which was admitted to the Naval Academy in 1860. They had hardly entered on their careers long enough to get the smell of the brine into their nostrils when the Civil War broke out. Here was the very chance they were longing for. But they ruefully saw two upper classes go out, and they knew that fighting of the larger sort was not yet for them.

For two years they were kept at their books, when finally the welcome news came that they would be graduated in three years instead of four, if they could pass the examinations. In spite of their many disappointments, there was a wild whoop of joy up and down the corridors, and they set about their work in earnest, studying with a concentration which no diversion could dissipate.

Taylor and Evans both left the Academy before having been graduated, and were ordered to duty with the blockading squadrons along the Gulf and Southern coasts. They went to their ships gleefully, bearing the proud titles of "acting ensigns," but in reality merely midshipmen of three years' standing,—destined, however, to do the duties and have the responsibilities of men many years their seniors in theoretical and practical service.

HOW CAPTAIN EVANS SAVED HIS LEG

Evans was in both attacks on Fort Fisher, and in the second fight he was shot twice. The wounds were severe, and he was sent into hospital. His leg was shattered badly, and after examining it carefully the doctors told the young sufferer bluntly that they would be obliged to amputate it.

When they went out Evans made a resolution that his leg was not to be cut off. He came to the conclusion that he would rather quit right there than to go through life one-legged. It was his own leg anyhow, and nobody had a better right to decide the question than himself.

By some means he got hold of one of the big navy revolvers, and had it secreted under his pillow when the surgeons, with a blood-curdling array of knives and saws, made their appearance on the scene and began preparations to carry their threat into execution. But when the chief surgeon turned to the bed to examine the wounds he found himself looking into the black barrel of young Evans's navy revolver.

"Now, see here," said Evans, as the doctor

retired in some alarm; "I want that leg to stay on. I need it. I will get well with that where it is, or not at all, and that's the end of it. That leg does not come off. Do you understand what I mean?"

The doctor was dismayed. But he understood perfectly, and Evans carried the day. The wounds were dressed and healed rapidly. In several months he was out again. But he limped then, and will as long as he lives.

SIGSBEE AS A PRACTICAL JOKER

Charles D. Sigsbee, writer, artist, hydrographic expert, mathematician, inventor, and incidentally the central figure, composed and dignified, in the greatest marine tragedy of modern times, is the kind of a man most people—men, women, and children,—like to see and know. His brow can be stern, and no one knows that better than the people who have sailed under him; but he loves peace better than war, and the twinkle behind his glasses never quite dies out.

As a midshipman he was always the prime mover in any affair which could contribute to the gayety of existence; was a better judge of people than he was of test-tubes, and a practical joker of ability, which is saying much.

The fascination which the ocean holds for all boys of sound mind gained an early sway over young Sigsbee. He received his appointment to the Naval Academy just before the Civil War, in 1859.

He liked the practical work, but could never settle down to the desperate grind of the academic course. He found himself more often making caricatures of "Dom Roget," the teacher of Spanish (a language he has since mastered), than in poring over the verbs and adverbs in the text-book. They were good caricatures, too, and when the other youngsters in the section saw them there was merriment which poor Dom Roget could not understand. But the professor solved the matter satisfactorily by marking all the delinquents on a low scale of credit, and, to be certain of the right culprit, Sigsbee lowest of all.

The young artist used to make pictures of everything and everybody he saw, and write pieces about them,—sprightly literature which went from one end of the Academy to the other. And so when the end of the year came round he found that, instead of being enrolled on the academic scroll of fame, he was relegated to the lower half of the class, which they called the "wooden" half.

He went back into the next class,—which

entered in 1860,—and with the advantage of a year of experience he obtained a position in the new class which he held until graduation time. He never quite got over his propensities for making fun.

He began as a joke, and afterwards kept it up, an anonymous correspondence with a member of his class. Sigsbee disguised his hand, and in the guise of "Lily Gaines," a very fascinating young woman of susceptible tendencies, wrote to Midshipman Mullan in such endearing terms that for three months that young gentleman was kept in a state of alternate suspense and rapture. At last, in a burst of confidence some one told Mullan of the deception, and the correspondence suddenly ceased.

But in spite of all this Midshipman Sigsbee went out into the world to practise his profession in stirring times, and ever acquitted himself as a valiant officer and accomplished gentleman. As the months rolled into years the naval service could boast of no officer who studied harder or who brought more steadfast qualities into his work.

THE BRAVE COMMANDER OF THE "WINSLOW"

Lieutenant John B. Bernadou was the commander of the "Winslow" in the fight at Cardenas, at which Ensign Worth Bagley, his second in command, was killed. The story of the fight these young officers made, until Bagley was killed, Bernadou was wounded, and the "Hudson" came and towed them out of danger, has been told again and again, and the tale of it will go down into the history of the Spanish-American War as one of the pluckiest of which there is record. Bagley, being the only naval officer killed during the war, was heard of from one end of the country to the other, but little was told of Bernadou, his commander.

Bernadou's early career showed in several instances the fearlessness of his disposition and the sturdiness of his character. The boy's first idea was to go to West Point. Failing in this, he secured an appointment to the Naval Academy, where he entered with a fine standing, which he maintained until he was graduated. He was always a brilliant worker, and in gunnery and foreign languages showed a most remarkable aptitude. To-day he speaks eight languages, and is one of the foremost men in the navy as an authority on smokeless powder.

THE MAN WHO NEVER KNEW FEAR

Bernadou's classmates say that he fears nothing on earth or water. His fearlessness overcomes any consciousness of self.

OUR NATION'S NEW HEROES

One afternoon in October, 1881, the United States steamer "Kearsarge," Captain G. B. White, lay at anchor in Hampton Roads. The weather had been stormy for a day or two, and the wind had kicked up a heavy sea. There was a strong tide running, and the vessel swung out on a long cable. A seaman by the name of Christoverson, who was boat-tender in one of the cutters swinging at the lower booms, went out and down the Jacob's ladder. In stepping to the thwart his foot slipped, and those on deck saw him disappear under the gray water.

There was a hoarse cry of "man overboard." Seaman Robert Sweeny, who saw the accident, running out along the boom, plunged in without delay, just as the man came up the second time. Bernadou, then a cadet-midshipman, heard the cry, and rushing to the gangway, saw the terrible struggle of Sweeny with the drowning man as the tide swept them out towards the sea. Bernadou tossed off his coat, and was overboard in an instant. Christoverson, in his fierce struggle, carried Sweeny down with him, the latter only breaking away to be carried down again.

Bernadou by this time was within reach, and catching the drowning man from behind, managed to relieve Sweeny until a line was thrown to them, and they were finally hauled aboard in an exhausted condition. For this act both Bernadou and the sailor received the recommendations of their captain and the thanks of William H. Hunt, then the secretary of the navy.

ONLY NAVAL OFFICER KILLED IN THE WAR

Worth Bagley's career at the Naval Academy was a triumph of the heart rather than of the mind. While he loved the service and hoped some day to fill a useful place in it, he found more to attract him in football and athletics than in calculus and least squares. But no man who ever entered was more beloved than he, and no man had better friends in the service and out of it. He was turned back twice, but entered, in 1891, the class of '95, in which year he was graduated. He was a member of the "Five B's," composed of Bennett, Barnes, Bagley, Breckinridge, and Baldwin, men who were close friends while they were at the Academy.

But football was Bagley's ruling passion. During this time, too, the great series of games between West Point and Annapolis, between the army and navy, over which the entire United Service went mad, were played, and Bagley was on the victorious team of '93, and was named for the "All-America" team.

Bagley roomed during the four years' course with his chum Breckinridge, who was washed off another torpedo-boat, the "Cushing," and drowned, as he was trying to get into Havana a few days before the blowing up of the "Maine."

"Worthless" Bagley (as his intimates called him) and Breckinridge were never left much to themselves in their quarters, for their room was always crowded during recreation-hours with cadets skylarking or asking advice or assistance. There was another intimate and classmate of Bagley, D. R. Merritt, who was killed in the "Maine" disaster a few days after the drowning of Breckinridge.

ROOSEVELT SAVED BAGLEY FOR THE NAVY

When Bagley came up for graduation at the end of the four-years' course the doctors thought they discovered an irregular movement of the heart, and recommended that he be dropped. Bagley took his case to Theodore Roosevelt, then assistant secretary of the navy.

Roosevelt, looking at him through his glasses with a quick, critical glance, said,—

"You are Bagley, the football player, are you not?"

Bagley said he was.

"Well, you are to stay in the navy while I am here. The service needs more men just like you."

Then Bagley went on his two-years' cruise, and when he came back he was passed through without question.

Captain Cook, Admiral Schley's chief-of-staff on the "Brooklyn," Captain Clark, of the "Oregon," and Commander Davis were room-mates in the famous class of Crowninshield, Taylor, and Evans. The "Brooklyn" and the "Oregon," commanded by classmates and room-mates, fought almost side by side down the desperate flight to the westward, the "Oregon" farther inland, but both thundering their iron missiles on the "Colon" as she struggled to her doom.

It is an interesting fact that Captain Clark, then holding the title of acting ensign, but really a midshipman, was the first one to communicate with the captain of the ram "Tennessee" when she was captured at Mobile Bay, while it was Captain Cook who received the surrender of the "Cristobal Colon." The third member of this trio was retired several years ago or he would have had a command in the same action. The affection which these youngsters bore one for the other was very much like that which existed between Captains Evans and Taylor.

CLARK'S HEROISM AT THE BATTLE OF MOBILE BAY

In the battle of Mobile Bay young Clark was on the forecastle of the "Ossipee," then holding an important position in the line of ships that swung past the torpedoes after the gallant Farragut in the "Hartford."

The forecastle was bare of any defence, and the position was exposed to all assaults of the fire, first from Fort Morgan and then from Fort Gaines, farther up. When the forts were passed, there followed a fierce fight with the gunboats and the invincible ram "Tennessee." Again and again the "Hartford," "Ossipee," and other vessels of the fleet rammed her in succession, and young Clark saw her terrible ports fly open and send out just by him their awful discharge.

At last, however, she became unmanageable, her shutters were jammed, and the "Ossipee," under full head of steam, was making for her. But while Clark was straining his eyes through the smoke, a white flag was hoisted in token of surrender. Clark shouted to Johnson, the commander of the ram, to starboard his helm. But the reply came that his wheel-ropes were shot away. It was too late to keep from striking her, but the force of the blow was broken by the manœuvre. This early experience was followed by the bombardment of Fort Morgan,—two important actions before Clark had got into his early twenties. His fearlessness then, as now, needs no mention.

POPULARITY OF CAPTAIN PHILIP AS A CADET

It has been said that Captain Philip's public acknowledgment of God on the decks of the battle-ship "Texas," after the fight before Santiago, was the natural expression of a deeply religious nature. But his classmates at the Naval Academy and the men who have sailed with him say that he is not more religious than other men in the navy,—not so religious as many, who always have their Bible on the table in their cabins and read it regularly when at sea or in port.

They believe that he spoke on the impulse of the moment, his heart devoutly thankful that the victory had been achieved at so slight a loss, and willing that all men should witness his profession of faith.

As a boy at the Academy, while he never surreptitiously drank, as others did, he made no pretence of being religious. He smoked

HER LAST DUTY

whenever he got a chance, in his quarters or in the darknesses back of old Fort Severn, between the watchmen's rounds. He never, as other cadets did, gave his word not to smoke, and so he felt a perfect freedom to do it if he could keep from being caught. Like Sigsbee, he was a practical joker, and if you should go to any of the members of his class and ask them who was the most popular man in it, they would say, "Jack Philip."

THE VERSATILITY OF ADMIRAL SAMPSON

In Admiral Sampson, the boy was father to the man. From boyhood his was a life of unneglected opportunities. Born of very humble parents, by the hardest of work and the most sincere endeavors he succeeded in obtaining his appointment to the Naval School. His mind, naturally studious, turned to the beginnings of the new profession with avidity, and so fine was his mind even then that, without trying himself unduly, he easily distanced his entire class and took first honors for the course.

His classmates say that he was studious, but they do not say that he applied himself so closely to the work that he shut himself off from the diversions or recreations of the rest-hours. On the contrary, he was foremost in most of the sports of the day, and was, in his own way, one of the best athletes in his class.

He was then, as he is now, an "Admirable Crichton," but his versatility did not diminish for him the serious aspect of any of the things he attempted. Some of his classmates called him cold, as his contemporaries out in the service do now, but when they wanted advice on any subject which seemed to require a reasoning power entirely beyond their own, they said, "Ask Sampson." He was not only high in his class councils, but dearly beloved, as he is to-day, by every man in it and every man who knew him. If people thought him cold then it was because they did not understand him. If they think him cold to-day it is because he does not care to be understood by the men with whom he has no interest or sympathy. If arrogance begins to be a virtue, then repression born of modesty is a crime.

To those men he cares for—now as in his youth—he has always a warm handshake and an open heart. His eye is calm, sympathetic, penetrating, stern, as the humor dictates, anything you please,—sometimes cold, but always hypnotic. If he wants the friendship of man or woman he is irresistible. To-day he is the authority on naval ordnance, an expert on explosives, a capital seaman, a famous tennis-player,—the best-equipped man in the service for any work—or play—that can be put before him.

BLUE, WHO DISCOVERED CERVERA'S FLEET

Victor Blue, who in his uniform made the fearless expedition ashore at Santiago, and actually saw for the first time the Spanish fleet within the harbor, is the kind of a man who does not have very much to say for himself, which is often a sign that a person is to be found ready when wanted. He was a member of the class of '87, in which his work was fair, but not remarkable in any way. He lived quietly, receiving his quota of good and bad marks, but having no special distinction, even in his offences against the oracles of Stribling Row.

He did not care much for "fems" (girls, in the vernacular), but towards his first class-year began to "take notice." He played a guard on the "Hustlers," the scrub football team which struggles with the "Academy" eleven on practice-days, but never made the "Team." He had plenty of grit, but was too light for the centre and not active enough for the ends. Blue is a fair specimen of the type of men who without ostentation have made our new navy what it is. Many men envy him, but no man begrudges him his numbers recently awarded for "extraordinary heroism."

YOUNG DEWEY AS A FIGHTER

George Dewey entered the class of '58 at the Naval Academy at the age of seventeen. He was not a large boy, but fairly up to middle height, and strong and active in all athletic sports. It was not long after his entrance that he found an opportunity to show the fighting spirit that was in him. It was not altogether of his own seeking, but when he was weighed in the balance, even then he was not found wanting.

The line between the Northern boys and the Southerners was clearly marked, and one day one of the Southerners called the young Vermonter a "dough-face."

Young Dewey awaited a favorable opportunity, and struck his opponent so fair a blow that he knocked him down. There was a rough-and-tumble fight then and there, and Dewey's adversary came out second best.

Later on another one of the Southerners insulted the young admiral, and there was another battle. But full satisfaction could not be obtained in this prosaic fashion, so the Southerner finally challenged young Dewey. The offer was promptly accepted, seconds were chosen, and the time and place were definitely settled upon. But some of Dewey's classmates, seriously alarmed at the aspect of affairs, and knowing that neither one of the principals was of a temper to falter, hastily informed the academic authorities, and the whole affair was nipped in the bud but a few hours before the hour set.

Dewey was graduated in 1858, and stood fifth in his class. Of the sixty-five who had started in as candidates, but fourteen received their diplomas at the end of the four years' course.

THE UNRECOGNIZED HEROES OF THE WAR

Much has been said and written of the heroes of action and movement. The country from one end to the other has rung with their praises. But what of the unknown heroes, unhonored and unsung? What of the men who, because of their superior abilities in other lines, were doomed to physical inaction? who performed their secret missions and labors skilfully, faithfully, uncomplainingly, while their classmates were being given numbers over their heads, and the chance of a lifetime for great deeds was being quietly passed by?

THE REAL BRAINS OF THE WAR.

Captain A. S. Crowninshield, the Chief of the Bureau of Navigation, bore the brunt of the brain-work for the men and ships at the front.

His bureau has to do with the ordering of all ships and all men, and Crowninshield, when he accepted the office, knew that the odds were against him. He knew that by his own orders he would put forward above him men who were many years his juniors in the service. He never winced, but went on perfecting the target-scores of the men behind the guns. When war was declared, he felt that, gun for gun, our navy could whip anything afloat. But he did not get out of the office. He could have had any command in Sampson's fleet. But he preferred to stay and carry out the work he had begun, in spite of the fact that each week, as younger men went over him, he saw the chances of hoisting the pennant of a fleet-commander grow fainter and fainter.

If you were to ask Secretary Long who did the real brain-work of the war, he would unhesitatingly answer, "Captain Crowninshield." Ask the younger officers in command of gun-divisions who is responsible for the straight shooting of the gun-captains, and they will say, "Captain Crowninshield." Ask any captain of the fleet of victorious battle-ships and cruisers of Santiago or Cavite who contributed most to the victory of Santiago and Manila, and they will say, "Captain Crowninshield."

These are the facts, and no one in the service disputes them for a moment. If the people are in ignorance, it is because Captain Crowninshield will never talk of himself or his own affairs under any circumstances.

Captain Crowninshield comes of a distinguished New England family. He is a grandson of Jacob Crowninshield, an early secretary of the navy, and a great-nephew of Benjamin Crowninshield, also a secretary of the navy. Like all the Crowninshields of Salem, he was full of love of the sea. His father was a graduate of Harvard and a founder of the Porcelain Club.

FILLING THE DUKE'S SHOES WITH MUCILAGE

Captain Crowninshield as a lad read and studied all the books he could find about the sea, upon which his ancestors, near and remote, had sailed. From the first he was determined to be a naval officer. To this end he went to a village where lived a member of Congress, who, he thought, might make him his appointee. The young man found the old member of Congress out in his field, ploughing. He liked the looks of the boy and gave him a half-promise of the appointment. Young Crowninshield was forced to wait a month, but at last the letter came, and with trembling fingers he broke the seal of the letter which made him a midshipman (a title which it is to be hoped will be restored ere long to the service).

Some of his classmates were the present Captain Clark, of "Oregon" fame, Captain Harry Taylor, Drayton Cassell, Captain Wadleigh, and Captain Cook, of the "Brooklyn." His room-mate was Pierre d'Orleans, and many a time did Captain Crowninshield rescue the young foreigner when the jokes became too fast and furious. A favorite amusement with the midshipmen was to fill "Pete" d'Orleans's shoes with mucilage. This practice, so far from making him feel like sticking to this country, persuaded the young duke to return to his native land, where there were no wild American boys to tamper with his dignity.

When the Academy was removed from Annapolis to Newport, young Crowninshield, of course, went with the school, with Evans and the others. He was told that those who could pass the required examination at the end of three years could go out to the war as officers.

Half of the class passed the examination. When one considers that no studying at night was allowed, that an officer made the rounds after lights were supposed to be out, and that at the sound of his footsteps the delinquent who was burning the midnight oil would be obliged to tumble into bed with his clothes on, throwing the wet towel which bound his head into the corner of the room, feigning sleep while a candle was passed across his face, one can understand why more young men of that class did not graduate at the end of the three-years' limit.

SCOUTING IN THE ENEMY'S COUNTRY

There are many other gallant navy men of whom the public has not heard, but two more will suffice. Within a week after the declaration of war two young ensigns, Ward and Buck, the former in the Bureau of Navigation and the latter at the Naval Academy, disappeared from the face of the earth. So completely did they destroy all traces of themselves that for all the Bureau of Navigation or their relatives seemed to know they might have ceased to exist.

Speculation was rife concerning them, but

nothing could be learned of their duties, the impression being, even among Navy Department officials, that they were installing a system of coast-signals in New England. Ward, it appears, disguised himself as an Englishman, and went straight into the heart of the enemy's country, making his headquarters at Cadiz, the principal Spanish naval station, and from there sending the Navy Department continuous and accurate reports of the fighting strength and actual movements of the Spanish fleet.

He was under suspicion, but watched his time, and succeeded in getting away to Porto Rico. There he was arrested as a suspicious character and spy. He managed, it is supposed through the British representatives, to obtain his release, and, escaping from San Juan, cabled the department a full account of the state of defences there and the movements of Cervera's fleet. While Ward was in Porto Rico, Buck was following Camara's fleet in the Mediterranean, keeping watch on its movements, and sending daily reports of its condition, armament, and plans.

We do not know what is in the hearts of men. We do not know whether the men who did the creditable things during the war did them in spite of themselves, or whether in the glory of action and adventure they took their lives into their hands gladly, fearlessly, for their country. We do know that there were hundreds ready and willing to court danger and death for a useful end who for lack of opportunity could not.

HEROES OF THE DEEP

ALL the long winter the "Polly J." had slept snugly in Gloucester Harbor, rigging unrove and everything snug aloft that the wind could freeze or the ice could chafe. Careful eyes had watched her as she swung at her moorings, and rugged hands had gripped the familiar gear as the skipper or some of the men had made their periodical visits. But however gray and desolate she loomed, with her topmasts housed and the black lines of ratline and stay across the brightening sky, nothing could hide the saucy cut-under of the bow and the long, free sweep of the rail.

The afternoon sun of March melted the snow on the south slopes of the fish-sheds, and great gray-and-green patches came out here and there against the endless white.

A brisk breeze, with a touch of the spring, blew up from the south, and the "Polly," heedless of the tide, turned her head to it, sniffing and breathing it, bobbing and jerking nervously at her anchor, impatient to be dressed in her cloud of canvas, and away where the wind blows free and the curl dashes high under the forefoot.

WHEN THE SNOW MELTS

Ashore in Gloucester town there are signs a-plenty of the work to come. The sleepy village throws off her white mantle and rises from the lethargy of the winter past. The spring is in the air, and the docks and wharves, white and ice-trussed during the long, bleak winter, are trod by groups of men, rubber-coated and "sou' westered," moving briskly from one shed to another.

In the town they gather like the stray birds of spring that flutter under the eaves of the store-houses. By twos and threes they appear. On street corners they meet, pipe-smoking, reminiscent, gloomily hopeful for the future, and grateful that they have helped themselves over "March Hill" without a loan from owner or buyer. And as they lounge from post-office to store, from store to shed, and back again, their talk is of dealings with owners and skippers, of vessels and luck.

For luck is their fortune. It means larger profits by shares, new dresses for the wife and little ones, and perhaps an easy time of it in the winter to follow. It means that there will be no long, hard winter of it at the haddock-fisheries at "George's," where trawls are to be set in weather which makes frozen hands and feet, and perhaps a grave in an icy sea, where thousands have gone before.

The skipper of the "Polly," even before he gets his men, has broken out his gear and reckoned up his necessities for the run up to the Banks. If he ships the same crew he had the year before, they work in well together. The "Polly's" topmasts are run up with a hearty will and a rush. There is a cheerful clatter of block and tackle, and the joyous "Yeo-ho" echoes from one schooner to another as sail and rigging are fitted and run into place.

The snow yet lingers in little patches on the moors when some of the vessels warp down to an anchorage. Dories are broken from their nests and skim lightly across the harbor, now alive with a fleet in miniature. Jests and greetings fill the air, as old shipmates and dory-mates meet again,—Gloucester men some of them, but more often Swedes, Portuguese, and men from the South.

For to-day the fleet is not owned in the villages, and Gloucester, once the centre of the fishing aristocracy, the capital of the nation of the Banks, is now but a trading- and meeting-place for half the sea-people who come from the North and East.

The skipper of the "Polly J.," himself perhaps the scion of three generations of fishing captains, may wag his head regretfully, for fishers cannot be choosers; but he knows that his fishing has to be done, and, after all, a "Portygee" is as good a sailor-man and dory-mate as another,—better sometimes,—if he keeps sober.

So long as the ship-owner makes his credit good at the store for the people at home, the fisherman takes life as joyfully as a man may who looks at death with every turn of the glass. If he takes his pleasures seriously, it is because he lives face to face with his Maker. Nature, in the awful moods he knows her, makes trivial the little ills that flesh is heir to.

So when the crews are aboard, and the stores and salt are being hoisted in, there is a hurry to be among the first away. Chains and windlasses creak and clang, nimble feet fly aloft, hoarse voices ring across the rippling water, and many a cheerful song echoes from ship to shore and back again.

Willing hands, strangers for months to hemp and tar, lay on to the tackle, as spar and boom are run into place. The fish-bins below are cleaned and scrubbed to the very quick.

Bright-work, if there be any, is polished, and sail-patching and dory-painting and caulking are the order of the day, and most of the night. The black cook, below in the mysterious blackness of the galley, potters with saucepan and kettle, and when the provisions are aboard serves the first meal. There is coffee, steaming hot in the early hours of the morning, and biscuit and meat,—plenty of it. There is not much variety, but, with the work to be done above and below decks, a full-blooded appetite leaves no chance for grumbling.

At last the bag and baggage of the crew are tossed aboard,—packs of tobacco innumerable, new rubber clothes, all yellow and shiny in the morning dampness, boots and woollens to keep out the cold of spring on the Bank Sea, —all bought on credit at the store, to be charged against "settling-day."

WAVING GODSPEED TO THE FISHER-FOLK

It is morning, just before the dawn. The "Polly J.," her new paint all silver in the early light, rides proudly at her anchor in the centre of the tideway. The nip of winter lingers in the air, but the snow is gone and the rigging is no longer stiff to the touch.

It is just daylight when the last dory is hoisted aboard into its nest. Three or four figures on the wharves, outlined against the purple sky and hills, stand waving Godspeed to their fisher-folk. Women's voices ring out between the creakings of the blocks, "Good luck! Good luck! 'Polly J.'; wet your salt first, 'Polly J.'" It is the well-wishing from the hearts of women, who go back to weep in silence. Which one of them is to make her sacrifice to the god of winds and storms?

There is a cheerful answer from the "Polly," drowned in the flapping of the sails and creaking of the windlass. The anchor, rusty and weed-hung, is broken out and comes to the surface with a rush, while sheets are hauled aft, and, catching the morning breeze, the head of the schooner pays off towards Norman's Woe, the water rippling merrily along her sides.

The figures on the wharves are mere gray patches in the mass of town and hills. The big sails, looming dark in the gray mists of the morning, round out to the freshening wind, and push the light fabric through the opal waves with ever-increasing speed. By the time the first rays of the rising sun have gilded the quivering gaff of the main, Eastern Point is left far astern, and the nose of the vessel ploughs boldly out to sea, rising with her empty bins light as a feather to the big, heavy swell that comes rolling in, to break in a steady roar on the brown rocks to leeward.

There is man's work and plenty of it during those sailing days past "George's," Sable Island, and the St. Lawrence. The provisions and salt are to be stowed and restowed, ballast is to be shifted, sails to be made stronger and more strong, fish-bins to be prepared, old dories to be made seaworthy, rigging to be tautened, and reels and lines to be cleared and hooked. Buoy-lines and dory-roding are to be spliced, and miscellaneous carpenter work takes up the time about the decks. For a skipper unprepared to take advantage of all that luck may throw in his way does an injustice to his owner and his crew. But, busy as the time is, the skipper has his weather-eye open for the "signs." The feel of the air, the look and color of the cold, gray Bank Sea, tell him in so many words how and where the fish will be running. At last a hand takes the heavy sea-lead and moves forward where the line may run free. Deliberately the line is coiled in great turns around the left hand, and then, like a big pendulum, the weight begins to swing with the strong right arm.

IN THE EXCITEMENT OF THE FIRST CATCH

There is a swirl of the line as the lead goes all the way over, a splash forward, and, as the skipper luffs her up into it, the line comes upright, and gets a depth of thirty fathoms. As she comes up into the wind, the noisy jib flaps down with a run, and the anchor drops to the sandy bottom. Now the buckets of bait are tossed up from below, and the skipper leaves his helm to take to the lines. Over the sides and stern they go, dragging down to leeward.

There is quiet for a moment, and then a line runs out. There is a tug as the strong arm checks it and hauls it in quickly, hand over hand. There is a gleam of light, a swish at the surface, and the fish flies over the rail, flopping helplessly on to the deck, the first catch of the season,—a big one.

Another tug, and another, and soon the work is fast and furious. It takes honest elbow-muscle, too, to haul ten pounds of floundering cod up five feet of freeboard to the rail and deck. Soon the deck is covered with the long, slim, gleaming bodies, and the boys of the schooner have man's work in tossing them into the gurry-pen amidships. Before the pen is filled, the fishes stop biting as suddenly as they struck on, and there is a rest for a while to bait-up and clean down.

If the signs hold good, the skipper will order the men out to set trawls, for the smell of the dead fish sometimes drives the school away.

HANDLING THE TRAWLS

The "trawls" are only an elaboration of the hand-lines. They are single lines, several hundred feet in length, with short lines and bated hooks at intervals. They are taken out by members of the crew in their dories, buoyed and anchored. It is the work of tending these trawls that takes the greatest skill and fearlessness. It is in the work of hauling and baiting the lines in all weathers that the greatest losses of life occur. There is no room on the decks of the schooners for heavy boats, and as many such craft are needed, five or six are piled together amidships. A block and purchase from aloft are their hoisting-tackle.

They are handy boats, though light, and two men and a load of fish can weather the rough seas, if your fisherman is an adept with his oars. But they are mere cockle-shells at the best, and are tossed like feathers. The "codders" are reckless fellows, and they will put out to the trawls day after day in any kind of weather, fog or clear, wind or calm, with not even a beaker of water or a piece of pilot-bread.

A LONELY NIGHT ON THE BROAD ATLANTIC

A night alone on the broad Atlantic in an open dory seems to have no terrors for them. Each year adds its lists of casualties to those that have gone before. Fogs have shut in, seas have risen, and morning has dawned again and again with no sign of the missing men. Sometimes an upturned dory is found, with her name—the "Molly S.," or the "Betty T.," in honor of the owner's shore-mate—on her pointed bow, but only the gray ocean can tell the story of the missing men.

When the "Polly's" day's luck is run, all hands take stations for dressing down. It is the dirty part of the business; but so quickly is it done that the crew seems part of a mechanism, working like clockwork. Two men stand at the gurry-pen, their long knives gleaming red in the sunset. The fish is slit from throat to tail with one cut, and again on both sides of the neck. It then passes to the next man, who with a scoop of his hand drops the cod's liver in a basket and sends the head and offal flying. The fish slides across the dressing-table, where the backbone is torn out by the third man, who throws it, finally, headless, cleaned, and open, into the washing-tub.

The moment the tub is filled, the fish are pitched down the open hatch to the fifth man, who packs them with salt snugly in the bins. So quickly is the work done that the fish seem to travel from one hand to another as though they were alive, and a large gurry-pen is emptied and the bin packed and salted in less than an hour.

WHEN THE DAY'S WORK IS DONE

The head of the black cook appears above the hatch-combing, and his mouth opens wide as he gives the welcome supper call. Down the ladder into the cuddy they tumble, one and all, and lay-to with an appetite and vigor which speaks of good digestive organs. Conversation is omitted. Coffee, pork-and-beans, biscuit,—nectar and ambrosia,—vanish from the tin dishes, until the cook comes in with the sixth pot of steaming coffee.

At last, when the cook vows the day's allowance is eaten and the last drop of coffee is poured, the benches are pushed back, tobacco and pipes are produced from the sacred recesses of the bunks, and six men are puffing out the blue smoke as though their lives depended on it.

The schooner rolls to the long ground-swell, her lamp-bracket swinging through a great arc and casting long, black shadows, monstrous presentiments of the smokers, which move rapidly from side to side over the misty beams and bulkheads like gnomes. A concertina, a mouth-organ, and perhaps a fiddle, are brought out, and a sea-song, an Irish jig, or something in unspeakable Portuguese, rises above the creaking of the timbers and the burst of foam alongside.

But the work is not done yet. It is never done. The ship is to be cleaned down and the gurry-pen and dory are to be sluiced out in readiness for the morrow. A vigil is to be kept, watch and watch, and woe be to the youngster who tumbles off his hatchway to the deck from sheer weariness.

WHEN A STEAMER LOOMS UP IN THE FOG

If there should be a fog,—and hardly a day or a night passes without one,—the danger is great. When the white veil settles down over the schooners the men on deck can hardly see their cross-trees. Foot-power horns are blown, the ship's bell is tolled steadily, while conch shells bellow their resonant note from the trawlers in the dories. But it is all to no purpose. For the great siren comes nearer and nearer every second, and the pounding of the waves against the great hulk and the rush of resisting water grow horribly distinct.

There is a hazy glimmer of a row of lights, a roar and a splutter of steam, a shock and the inrush of the great volume of water, a shout or two from the towering decks and bridge, and the great body dashes by disdainfully, speed undiminished, her passengers careless, and unmindful that the lives and fortunes of half a dozen human beings have hung for a moment in the balance of Life and Death. But records have to be made, and the gold-laced officers forget to mention the occurrence. The men on the schooner do not forget it, though. More than one face is white with the nearness to calamity.

"What was she, Jim?"

"The 'Frederick.' I'd know her bloomin' bellow in a thousand."

They lean out over the rail and peer into the gray blackness, shaking their fists at the place where she vanished in the fog.

The man who gets his name in the newspaper and a medal from his government is not the only hero. And the modesty with which the Gloucester fisherman hides his sterling merit is only convincing proof of the fact, —Gloucester is a city of heroes.

For grit and devotion the case of Howard Blackburn surpasses understanding.

THE COURAGE OF THE UNNAMED HEROES

Blackburn and his dory-mate left their schooner in a driving snow-storm. Before they had been at the trawls long the weather had become so thick that they couldn't see ten feet from the dory's gunwale. The wind shifted and put them to leeward of their vessel. There was never a sound of bell or horn through the thickness, and, though they pulled to windward, where they thought their skipper lay, the vessel could not be found. They were lost, and the sea was rising. Then they anchored until dawn.

When the snow stopped falling, they saw the schooner's light, a tiny speck, miles to windward. To reach it was impossible. The situation was desperate. Wave-crest after wave-crest swept into the dory, and all but swamped her. Time after time she was baled out, until it seemed as if human endurance could stand it no longer. Blackburn made a sea-anchor for a drag, but in throwing it out lost his mittens overboard. It was horrible enough to fear drowning in the icy sea, but as he felt his hands beginning to freeze the effort seemed hopeless.

With hands frozen, Blackburn felt that he was useless, for his dory-mate was already almost helpless with exposure. So he sat down to his oars and bent his freezing fingers over the handles, getting as firm a clutch as he could. There he sat patiently, calmly, keeping the dory up to the seas meanwhile,—waiting for his hands to freeze to the oars. The dory became covered with ice, and pieces of it knocked against the frozen hands and beat off a little finger and a part of one of the palms. During the second day Blackburn's dory-mate gave it up, and Blackburn laid down beside him to try and warm him. But it was useless. The dory-man froze to death where he lay.

FOR FIVE DAYS ADRIFT AND STARVING

When Blackburn felt the drowsiness coming over him, he stood up and baled as the boat filled. The third day dawned without a ray of hope, and not a morsel to eat or a drop to drink, so he stuck the oar through his wounded fingers and rowed again.

The fourth day he saw land. He did not reach it until the afternoon of the fifth day, when he landed at a deserted fish-wharf. No one could be found, and he was too weak to move farther. So he lay down, more dead than alive, and tried in vain to sleep, munching snow to quench his thirst.

The next day he went out in the dory to try to find some signs of life, and in about three hours, the last remnant of his strength being gone, he saw smoke and the roofs of some houses, and he knew that he was saved. Even when he reached the shore in a pitiable condition, he would not go into the house until they promised him to get the body of his dory-mate.

This heroic man lost his hands and the most of his toes, but he reached Gloucester alive. The story of his grit and devotion to his dory-mate are to-day told to the young fishermen of the fleet, and the men of the Banks will sing his praises until Time shall have wiped out all things which remain unrecorded.

WHERE THE COD ABOUND

On some of the schooners, by the middle of the season most of the salt is "wet." It is then that the "Polly J." follows the fleet up to the "Virgin."

This is a rocky ledge, many miles out in the desolate Bank seas, which rises to within a few feet of the surface of the ocean. Here the cod and camplin abound, and here, when it is time for them to run, most of the schooners come to anchor, sending out their little fleets until perhaps two thousand dories and schooners are afloat at the same time, within a distance of two or three miles of one another. When the schools of camplin come to the surface and begin to jump, the dories all close in on them, for the fishermen know that the cod are after them. Almost as quickly as the lines can be baited and cast overboard the fish strike on, and the work is steady and hard until the dories, loaded down almost to the gunwales, have made several trips of it, and the salt in the bins shows a prospect of being "all wet" before the week is out.

The few days towards the end of the season at the "Old Virgin" are a race between the ships at catching and dressing down. The rival crews work from dawn until dark.

At last the big mainsail of the victor—perhaps the "Polly J."—is hauled out, the chain is hove in short, and the dories from less fortunate schooners crowd alongside with good wishes and letters for the folks at home. Anchor up, the flag is hoisted,—the right of the first boat off the Banks,—and the proud schooner, low lying in the water with her fifteen hundred quintal, bows gracefully to each vessel of the fleet at anchor as she passes them, homeward bound.

HEROES OF THE DEEP

WHEN THE SCHOONERS MOVE UP THE HARBOR

Homeward bound!—there is magic in the word. Though the first vessel to head to the southward is proud among the fleet, she has a burden of responsibility upon her, for she carries every year news of death and calamity that will break the hearts of many down in Gloucester, and the flag she flaunts so gayly must come to half-mast before she sights the hazy blue of Eastern Point.

During those long summer months a lonely wife goes about her household duties down in Gloucester town. There is a weight upon her heart, and until the fleet comes in and she sees the familiar face at the front gate, happiness is not for her. Day after day she listens for his footsteps, and after supper, when the season draws to a close, she walks down to where she can look far out to sea.

Then a schooner, heavy laden, appears around the Point. She comes around and moves up the harbor slowly,—oh, so slowly. The flag the wife has seen is half-masted, and she knows that some woman's heart is to break. Will it be hers?

OUR NAVY IN THE GREAT WAR

NO volume, however fragmentary, on the achievements of our Navy would, of course, be complete without a summary of the deeds accomplished by its ships and men in the Great War in Europe. The fights of the old sailing frigates yard arm to yard arm with their enemies, the boarding parties with pikes and cutlasses were picturesque and developed incidents of fine personal daring. Perhaps in these particulars, the more modern story of achievement loses something in color. Deeds of daring were not lacking in the great European War but they were not always deeds of physical prowess, of muscular strength and skill, though these too were so common that many of them have passed into the logs of the ship unhonored and unsung. In most of the achievements of the Navy in the Great War, courage of a much higher type was required, a spiritual strength and endurance which enabled the crews of ships to "carry on" in the face of the deadly menace of the submarine, the torpedo and the hidden mine, agencies of sudden death more certain and terrible than the imperfect instruments of war of the days of our forefathers. It takes, I think, a finer courage to stand to one's duty and await such menaces than to join in the desperate rush of men in the heat of battle.

The great achievement of the United States Navy during the war was, of course, the transportation and convoy of the ships of troops and supplies to Europe—"the Great Atlantic Ferry Company" as it was called "Incorporated but Unlimited." To carry more than a million men across the Atlantic Ocean in the face of the waiting submarines of the enemy required a skill in organization and a perfection in plan and technique that are worthy of the highest praise. The very sight of these formidable armadas must have exercised a strong repressive influence on the ambitions of the submarine commanders who happened to be on the lookout for easy prey. For the transports were moved in two columns, *en echelon*, guarded on each side by trawlers, also *en echelon*, they in turn guarded on the outside and front and rear by Destroyers moving in a zig-zag course which combed the sea in all directions.

These Destroyers did noble work in the convoying and in the submarine hunting in the North Sea. Immediately after the signing of the War Resolution by President Wilson, the Navy Department ordered all destroyers on duty along our coast to fit out at once for distant service. Their hulls were scraped and painted and stores and provisions were quickly put aboard. When the fleet set out, no Commander knew where he was going, for, as a measure of security, the instructions were to proceed to a point fifty miles off Cape Cod and there open sealed instructions. These, the commanding officer found, directed them to proceed to Queenstown—an epoch making voyage, fortunately without casualties, though it would have been a fine feather in the cap of the Germans to have sunk at least one of the ships of this new enemy from across the water which was rushing to the assistance of its ally and brother. During the previous three weeks German submarines had sunk 152 British merchant ships. Upon these ships the English people were relying for their sustenance. The situation looked very dark indeed.

It may be imagined with what joy the British fleet hailed the arrival of the doughty

Americans. The British Admiral, Bayly, after the formal ceremony of greeting, asked the Americans:

"When will you be ready for business?"

And the reply, verified by later inspections, was both surprising and reassuring.

"We can start at once, sir."

In active work the Navy seems to have figured chiefly through its Destroyers. Submarine hunting was very popular with our Jackies and though the work was hard, the hours endless and unremitting vigilance the price of security, there was glory to be won and joy to be found in the very rigors of the game. Until the perfection of the depth bomb, the advantage lay always with the submarine, for the Destroyers were constantly exposed while only the periscopes of the submarines were visible. The submarines avoided the Destroyers when they could for they were in search of less poisonous foes. But when Destroyer met Submarine, there was swift and immediate action, the Destroyer heading at full speed toward the vanishing periscope, firing with all its batteries in the hope of damaging the enemy. Sometimes a line of bubbles on the surface of the ocean betrayed the whereabouts of a sunken foe, when the depth-charges on the Destroyer were let go. It was hit or miss, then, for though two or three hundred pounds of explosives were used and the reverberations could be felt for miles, no damage was done to the hull of the submarine unless the explosion occurred within one hundred feet of the target. How many submarines were put out of commission by American Destroyers can not, of course, be told with accuracy, but after the introduction of the depth bomb, the appearance of oil and debris upon the surface of the ocean usually told of the grim fate of the silent hunter of the deep.

The death blow to the submarine menace, bringing with it the submission of the forces on land and the quick ending of the war, was delivered by the imprisonment of the enemy's submarines within the confines of the North Sea. This was accomplished by the submarine mine barrage laid down across the northern end of the North Sea between Norway and the Orkney Islands. Efforts had been made by the British Navy to make the ingress and egress of the German submarines to and from their bases as dangerous and difficult as possible, but it remained for the Bureau of Ordnance of our Navy Department to perfect a plan so stupendous in magnitude that it was not believed possible of effective execution, a bold scheme, its critics said, foolish, impossible, to lay a mine field across a body of water 230 miles wide, or as long as the distance from New York to Boston. And yet, by unparalleled exertions in manufacture, and in intensive training and equipment, this astounding feat was accomplished.

A submarine mine consists, briefly, of a mine-case, shaped like a ball or egg about one yard in diameter and an anchor in the form of an iron box about two feet square, connected by a wire-rope mooring cable the size of one's little finger. The mine case contains the charge of high explosive, 300 pounds of T N T and the firing mechanism. Each of these mines with its attachments weighed about 1400 pounds. And in the tremendous feat, more than seventy thousand of these gigantic eggs, spawn of the diabolic necessity of war, were sown across the North Sea, to float harmless below the surface of the sea until touched by the prow or sides of a German submarine.

The mine fleet was under the charge of Captain Reginald R. Belknap, U.S.N., and four fifths of all the mines laid were sown by the American fleet. The motto of the fleet was "stick to your job and if necessary go up with it." There were countless volunteers for this work and those who were fortunate enough to be chosen to belong to the squadron felt themselves to be objects of envy. Hour after hour, week after week, the mining went on, the "U. S. S. Housatonic" gaining the record of planting a string of 675, one every 11½ seconds in a period of two hours and ten minutes. In a later excursion the "U. S. S. Canonicus" laid 860 mines in 3 hours and 35 minutes without a break, a string longer than the distance from Washington to Baltimore.

Periscopes were sighted and attacks were made upon the German submarines by watchful Destroyers. There were difficult, hazardous moments when imperfect mines exploded but the work went on to completion.

The actual losses to German submarines will probably never be known; but according to reports, the Germans admitted the loss of 23 submarines among these mine fields and the British Admiralty staff have been quoted as holding that the surrender of the German fleet and the final armistice were the result largely of the failure of the submarine warfare, this failure being admitted as soon as the mine barrage was found to be effective.

How was it possible, to navigate safely so

many times through infested waters and close to mine fields, to steam so many miles so steadily, to outdo the telephone in speed by signaling and to handle and plant all these mines—sometimes in fog and darkness—all without accident? The answer is—logical and thorough preparation, co-operation and able leadership, unquenchable optimism and courage.

To these fine achievements of the ships and personnel of the Navy itself must be added one glorious page in the history of the Marine Corps, which was written at Chateau Thierry where the American marines held the line against the most vicious wedge of the German Army, in its almost successful advance on Paris. With only 8,000 men engaged in the fiercest battles, the casualties of the Marine Corps numbered 69 officers and 1531 enlisted men killed and 78 officers and 2435 enlisted men wounded.

It was on June second, 1918, that the American Marines had their desperate call to battle and the honor of throwing back the Prussian Guard Regiments beginning the retreat which lasted until Germany was beaten to its knees in final surrender. The ex-Secretary of the Navy, Mr. Daniels, has described the battle in the following glowing words:

"The United States marines, trained to keen observation upon the rifle range, nearly every one of them wearing a marksman's medal or better, that of the sharpshooters or expert riflemen, did not wait for those grey-clad hordes to advance nearer. Calmly they set their sights and aimed with the same precision that they had shown upon the rifle ranges at Paris Island, Mare Island and Quantico. Incessantly their rifles cracked, and with their fire came the support of the artillery. The machine gun fire, incessant, also began to make its inroads upon the advancing forces. Closer and closer the shrapnell burst to its targets. Caught in a seething wave of machine-gun fire, of scattering shrapnel, of accurate rifle fire, the Germans found themselves in a position in which further advance could only mean absolute suicide. The lines hesitated. They stopped. They broke for cover, while the Marines raked the woods and ravines where they had taken refuge with machine gun and rifle to prevent their making another attempt to advance by infiltrating through. Above, a French airplane was checking up on the artillery fire. Surprised by the fact that men should deliberately set their sights, adjust their range and then fire deliberately at an advancing foe, each man picking his target instead of firing merely in the direction of the enemy, the aviator signaled below 'Bravo!' In the rear that word was echoed again and again. The German drive on Paris had been stopped."

There was fighting in Belleau Wood even more severe. Companies of Marines that had entered 250 strong dwindled to 50 and 60 with a sergeant in command but the constant attack did not falter. The fighting was from tree to tree in the American fashion, a rush, a halt, a rush again in four-man formation. Waves, more waves, the marines battle cry "E-e-e-e Y-a-a-h-h Yip!" resounding from one end of the woods to the other and spreading fear in the hearts of the German troops who saw in those "Devil Dogs" from America, the personification of desperate courage which did not know defeat.

At Soissons, the story was the same, at the St. Mihiel Salient, at Blanc Mont Ridge. In the words of General Lejeune of the Marines to his men, "To be able to say 'I belonged to the Second Division; I fought with it at the battle of Blanc Mont Ridge' will be the highest honor that can come to any man."

As to the "Grand Fleet," as it was called, the very backbone of the structure that was to bring a victorious peace, the work of British Admiral Sir David Beattie and under him, Admiral Hugh Rodman of the United States Navy, is worthy of the highest praise. The activities of the American vessels in conjunction with the British meant the elimination of petty jealousies, absolute co-operation and perfect team-work. Sometimes the American squadron was under the command of British officers and sometimes British officers served under Admiral Rodman; but there was never the slightest friction. It was the policy of the Grand Fleet to go after the enemy every time he showed his nose outside of port. Every inducement was offered him to emerge. Inferior forces were sent to lure him from his bases, valuable convoys were despatched, apparently without protection, but he would not come out. There were submarine attacks however and many lucky escapes managed by skillful manœuvering on the part of American ships. Add to the uncertainty of constant danger, the terrific weather of the North Sea in winter, the cold, sleet, snow and ice and heavy seas, the continuous cruising in close formations at high speeds and you will form a mental picture of the trials of the men of our Naval service during the final months of the war. But the Grand

Fleet accomplished its purpose at last and the German ships were surrendered at Scapa Flow, a bleak harbor in the Orkney Islands. The seas were made safe and free to the ships of the allies. If, in the last word, sea power really dominates the world, the navy of the United States has shown our brothers and our allies across the water that in a final reckoning upon the sea. it is well able to uphold the fine traditions of the past.